# I Never Call It Big Bang
## George Gamow
The Extraordinary Story of a Genius of Physics

# I Never Call It Big Bang
## George Gamow
### The Extraordinary Story of a Genius of Physics

**Alessandro Bottino**

**Cristina Favero**

 World Scientific

NEW JERSEY · LONDON · SINGAPORE · BEIJING · SHANGHAI · HONG KONG · TAIPEI · CHENNAI · TOKYO

*Published by*

World Scientific Publishing Co. Pte. Ltd.

5 Toh Tuck Link, Singapore 596224

*USA office:* 27 Warren Street, Suite 401-402, Hackensack, NJ 07601

*UK office:* 57 Shelton Street, Covent Garden, London WC2H 9HE

**Library of Congress Cataloging-in-Publication Data**

Names: Bottino, A. (Alessandro), author. | Favero, Cristina, author, translator.

Title: I never call it big bang : George Gamow : the extraordinary story of a genius of physics /
    Alessandro Bottino, Cristina Favero ; English translation by Cristina Favero.

Other titles: Non l'ho mai chiamato big bang. English

Description: New Jersey : World Scientific, [2022] | Translation of: Non l'ho mai chiamato big bang. |
    Includes bibliographical references and index.

Identifiers: LCCN 2021058931 (print) | LCCN 2021058932 (ebook) |
    ISBN 9789811242304 (hardback) | ISBN 9789811242311 (ebook for institutions) |
    ISBN 9789811242328 (ebook for individuals)

Subjects: LCSH: Gamow, George, 1904–1968. | Physicists--Biography. |
    Physics--History--20th century. | Big bang theory.

Classification: LCC QC16.G37 B6813 2022  (print) | LCC QC16.G37  (ebook) |
    DDC 539.7092 [B]--dc23/eng/20220125

LC record available at https://lccn.loc.gov/2021058931

LC ebook record available at https://lccn.loc.gov/2021058932

**British Library Cataloguing-in-Publication Data**

A catalogue record for this book is available from the British Library.

Originally published in Italian as *Non l'ho mai chiamato Big Bang*

© 2020 by Alessandro Bottino and Cristina Favero

English translation by Cristina Favero

For any available supplementary material, please visit
https://www.worldscientific.com/worldscibooks/10.1142/12420#t=suppl

Printed in Singapore

To Erik, Matilde and Gabriele

# Acknowledgments

We would like to thank the American Institute of Physics for the permission to use excerpts from Charles Weiner's interview to George Gamow [AIP] for this book.

We would also like to thank Eli Dwek, for allowing us to publish his cartoon about the cosmic thermostat.

# Contents

# First Clues about George Gamow

Have you ever stopped and watched cows as they graze fresh grass in the countryside? You probably have. But have you ever really observed the way cows ruminate? There is someone who decided to go deep into this matter and understand whether cows move their jaws in a horizontal direction, or just up and down, or maybe simultaneously in both directions, in a circular movement. And if the movement is actually rotational, is it clockwise or counter-clockwise? We bet that this is something only some eccentric physicists might decide to spend their time on.

In fact, in 1927 two young physicists, Pascual Jordan and Ralph de Laer Kronig, wrote a short paper for *Nature,* one of the most prestigious scientific journals in the world, with the following title: *Movements of the Lower Jaw of Cattle during Mastication* [JdL].[1] Shaping the article in a seemingly serious way, they stated that, according to their own observations, in cattle rumination clockwise rotation movement slightly prevails over counter-clockwise! Note that these two scientists were both working in Copenhagen, at the Institute for Theoretical Physics directed by Niels Bohr, one of the greatest experts in quantum mechanics, and were both engaged in important scientific research work.

George Gamow, the protagonist of our story, had obviously come across this paper, and having immediately grasped the subtle irony hidden in between the lines, he decided to build up his own funny story around it. He told his friends[2] that he had personally spent time in Denmark

---

[1] A symbol in square brackets denotes a reference in the *List of References* section.

[2] This anecdote is recalled by biologist Alexander Rich in the Reminiscences section of the proceedings of the *George Gamow Symposium*, held at the George Washington University in 1996 [HPA].

observing ruminants, and that he had noticed that their mastication was clockwise. And that he had then repeated the same observations during a stay in Brazil, where he had verified that the rotation was in the opposite sense. So, Gamow concluded, probably Coriolis forces[3] were involved. Gamow ended his joke stating that he had sent a letter about this discovery to *Nature*, but that, sadly, it had been rejected.

But who was George Gamow and why tell his story?

Gamow is the theoretical physicist to whom modern physics owes credit for many of the most important discoveries of the past century, especially during the period between the late 1920s and the 1960s — a period characterized by incredibly fast developments in physics.

He was born in Odessa (Ukraine) in 1904 and died in Boulder (Colorado, USA) in 1968. He lived across a very complex historical period, marked by conflicts (The First World War, The Russian Revolutions, The Second World War) and important political and economic events (Stalin's rise in the Soviet Union, the Great Depression, the ascent of the Third Reich in Germany and then the Cold War). Many of these events affected his life very deeply, especially his younger years in Russia. But when he chose to move to the United States of America in 1934, he somehow managed to avoid some of their most dramatic outcomes.

The first quarter of the 20th century had been marked by the birth of new fundamental theories in physics: quantum mechanics, special relativity, and general relativity with its implications for cosmology. As a consequence, by the end of the 1920s, a completely new theoretical framework, along with many ground-breaking experimental discoveries, opened up new wide fields of research.

George Gamow was one of the key figures of that time. His most remarkable achievements were the result of a combination of originality

---

[3] Coriolis forces are responsible for the opposite sense of rotation of air masses in cyclones in the northern and southern hemisphere of the Earth.

and daring in his approach to science, in a mix of expertise and natural talent that together define his profile as that of a true revolutionary scientist.

Gamow was the first physicist to apply the newly born theory of quantum mechanics to nuclear physics. In 1928, he dismantled the theory of radioactive alpha decays (based on classical physics) as suggested by Ernest Rutherford, with a straightforward interpretation of alpha decays in terms of an effect (the so-called *tunnel effect*), predicted by quantum mechanics. An impressive result for a newcomer, as he was.

More or less in the same years, he provided his experimental colleagues with the ideas that inspired the building of the first particle accelerator, which employed protons as particle projectiles and turned out to be suitable for investigating the atomic nucleus. Gamow also pioneered the idea that a nucleus can, under certain circumstances, be described as a liquid drop — a model that later on would become the keynote for the interpretation of important nuclear transformations, such as fission and fusion.

These were just some of the results of his brilliant early years, when he was in his mid-twenties. In the following years he completed, together with Edward Teller, the theory of beta decay put forward by Enrico Fermi.

More breakthroughs followed when Gamow moved his scientific interests from nuclear physics to other research fields. In astrophysics he established the quantitative basis for the evaluation of thermonuclear reactions in stars, and modelled a number of astrophysical objects, such as red giants and neutron stars. In this same field, he also suggested that the *neutrino*, Wolfgang Pauli's newly invented particle, was to be considered a main actor involved in supernovae explosions — and thus took the neutrino out of the laboratory benches where it was usually investigated within beta decay processes, and into the universe!

Gamow's flow of new intuitions moved further into cosmology where he, as the natural heir of Alexander Friedmann and Georges Lemaître (the fathers of the formulation of an expanding universe), developed the model of a dense and hot primordial universe — the model that is currently known as the *Big Bang Model* (although he personally always refused to use this *misleading* name, that had been sarcastically given to his model by Fred Hoyle, one of his main rivals in cosmology). Working on this model, by the end of the 1940s Gamow and his collaborators reached the conclusion that the lightest chemical elements in the universe must have been produced in a time lapse of a few minutes. This same theoretical framework led to the intriguing prediction of the existence of a remnant blackbody cosmic background radiation, of about 5 kelvin, a value amazingly close to the actual temperature measured by the COBE space mission in 1989!

In the 1950s, Gamow ventured into a field that was very far from his professional expertise, and tried to decipher the code according to which the genetic information contained in the DNA is transferred during the process of protein production — too daring an initiative, that inevitably caused indignation in part of the scientific community at the time.

Gamow was firmly convinced that scientific progress can be achieved only on two fundamental premises: sharing of results and collaboration. He himself, for the benefit of his own research, needed to share new intuitions and ideas with his co-workers, and whatever came to his mind had to constantly be confronted with his colleagues' opinions. In this spirit, he became a great promoter of the organization of meetings and cycles of conferences in the United States, inspired by what he had experienced at some European institutes. From the mid-1930s to the 1940s, this became a relevant part of Gamow's scientific engagement, with important repercussions for the entire scientific community.

Besides, as to his approach to research, Gamow was convinced that the various branches of physics should not proceed separately, but rather merge into forms of interdisciplinarity. A remarkable example of this were the applications of quantum nuclear physics to astrophysics, that practically gave birth, way before its formal recognition, to a brand-new field of research that more recently has been named *astroparticle physics*.

Gamow's approach to physics was absolutely unique. All his extraordinary scientific achievements came along with a sense of humor, often through jokes and tricks. To him, being a physicist meant, above all, having fun and amusing others. He surely was talented, and his natural self-confidence allowed him to show off his professional skills intertwined with funny jokes. Some of the tricks he played on his colleagues became memorable — we will mention some of them in our story.

George Gamow never passed unnoticed and impressed people from the very first encounter. This is how a member of the George Washington University in Washington D.C., where Gamow was a teacher and a researcher from 1934 to 1956, describes the impression he got the first time he saw him: "In August 1934 there appeared on the GW campus a 6-foot-3-inch, 30-year-old, flaxen-haired, Ukrainian émigré scientist. His startlingly blue eyes twinkled myopically behind lenses that resembled the bottoms of cider bottles. He conversed with a cosmopolitan circle of friends in a variety of European languages, with a fractured but poetic delivery that was animated and usually high-pitched. His name: George Gamow" [Har1].

Certainly, what was most striking about Gamow was his great culture and his outstanding intellectual abilities, that were combined with an incredible creativity: by mixing all these personal qualities together, he could face and solve problems in new and unpredictable ways.

His colleagues and students were fascinated by the great enthusiasm with which he used to announce a new result, as he also used to share all the fun he had experienced throughout the research.

He was impulsive and sometimes too superficial, and this often even led him to wrong or imprecise results. But Gamow claimed the right to make mistakes, as an unavoidable implication of carrying on research in unexplored directions and towards brand new ideas.

Gamow was also a remarkable science communicator. To popularize science was almost natural to him, something that he really enjoyed, and he carried on this activity in parallel with his research throughout his life. In 1928 he invented a character, Mr. C. G. H. Tompkins, an unlikely bank clerk who accidentally develops a passion for physics. He was to become the protagonist of a series of books about his incredible adventures in imaginary worlds. Through these stories, Gamow succeeded in making very complex topics, such as quantum mechanics and nuclear physics, interesting and understandable to a wide audience of lay readers.

The many books Gamow wrote about Mr. Tompkins soon became best sellers and were translated into many languages. All this, in spite of the many negative comments he received by academic colleagues, who blamed him for his informal style and, to put it in physicist (and science historian) Wolfgang Yourgrau's words: "We do not fancy the oversimplifying, popularizing of our science... it is tantamount to a cheapening of the sacred rituals of our profession... many of us considered him washed up, a has-been, an intemperate member of our holy order".[4]

Was Mr. Tompkins George Gamow's *alter ego*? Some clues make us believe so, especially when Tompkins's name appeared in the list of acknowledgments in a scientific paper, published with Edward Teller [GT39] in the famous scientific journal *Physical Review*, with official thanks for "having suggested the topic"!

Despite the many contributions of great scientific importance that Gamow gave to physics, astrophysics and cosmology, during his life he was not acknowledged with all the credits he actually deserved.

He was often criticized for being too impetuous, for elaborating ideas too roughly and for being lazy when it came to solving mathematical derivations. He was also blamed for taking shortcuts, even *quick*

---

[4] Quote reported by Eamon Harper [Har2] as excerpted from an article by Wolfgang Yourgrau [You].

*and dirty* ones, just to easily get to a result. In a way, he was criticized for behaving and acting according to his own nature! Or perhaps the academic world of his time simply could not cope with his uncontrolled and uncontrollable creativity.

There is quite a lot of information available about George Gamow. Our main source of information about his younger years is for sure his autobiography, that he wrote during the last years of his life. Its title is *My World Line. An Informal Autobiography* [GG70] and it is a very unusual book, as one might expect from this author. As Gamow states in the preface, his intention was to collect a series of personal stories that he would have loved to tell a group of friends after a nice dinner, sitting in front of the fireplace. *"My World Line"* is an expression borrowed from the theory of relativity, where it indicates a path in space-time, and Gamow uses it here to symbolize his own personal and scientific journey.

In his autobiography, Gamow describes in great detail, and often in quite a picturesque way, his personal and scientific evolution during the first half of his life. His youth in Russia and the crucial years he spent at some of the most important "sanctuaries" of the late 1920s physics, such as the Institute for Theoretical Physics in Göttingen, directed by Max Born, Niels Bohr's Institute for Theoretical Physics, situated at n.15 on Blegdamsvej in Copenhagen and the Cavendish Laboratory in Cambridge, where Ernest Rutherford was conducting his experimental research in nuclear physics. Gamow enriches the description of this period by remembering a series of anecdotes about his younger years — exactly as one would do when sharing stories with friends. In between the lines, one can clearly perceive a strong nostalgic feeling, especially when his memory goes back to his motherland, Russia, that he had permanently left in 1933 and had never gone back to.

About the many years spent in the United States, Gamow writes very little, and just shares some very short descriptions. Perhaps, he was planning to write more about this period later on, and died before he ever managed to. Still, according to his colleague and friend Stanislav Ulam (who wrote the preface of *My World Line*) Gamow probably avoided writing about his activity in the United States intentionally, in order not to hurt anyone with his words. So, most probably it was his clear choice.

Gamow's life after 1934 is well reconstructed in the work of Eamon Harper, a professor at George Washington University [Har2]. Harper is the author of a short but very interesting biography, focused on his scientific work. We have widely used information from it too, for our book.

However, our story is mainly based on the interview that Gamow gave to Charles Weiner in the spring of 1968. A transcript of this interview is kept at the American Institute of Physics, and is available on the web [AIP].

Some info about this interview:

*Place.* Gamow's home in Boulder, Colorado, where he moved to in 1956, after leaving Washington.

*Time.* It is a two days long interview, that took place on April 25 and 26, 1968. Gamow is already in quite bad health, due to alcoholism and heavy smoking and will die in August that same year.

*Participants.* Three people partake in the interview. George Gamow, the protagonist. The interviewer, Charles Weiner, who besides being a well-known science historian, has been a key figure in Oral History, a method of collecting data and information by means of direct interviews with scientists, rather than referring only to printed material. Weiner was also professor at the Massachusetts Institute of Technology and a careful analyst of political, social and ethical issues. The third participant to the interview is Gamow's second wife, Barbara Perkins, who was *publicity manager* for the editor that printed most of Gamow's popular books and widely contributed to their success. They got married in October 1958. She does not partake directly in the interview, but is present whenever it comes to remind Gamow of specific episodes or find papers and documents.

In his very professional interview, Weiner goes through all Gamow's major scientific results, consulting original papers and listening to anecdotes about his life. Their tone is very confidential. On the second day, the interview ends with a dinner at Gamow's house, and their conversation continues as Gamow keeps telling funny stories about his past and his colleagues.

The long report of the conversation with Gamow written by Weiner, is very interesting and represents an excellent example of scientific journalism, providing crucial information for understanding the complexity of both George Gamow's scientific career and personality. Stimulated by his interviewer, Gamow looks back at the most important ideas, events and people who influenced his professional and personal path, thus also providing a unique description of the development of physics in the past century. Gamow sometimes even expresses his personal opinion about facts and people, something he avoided doing in his autobiography.

This interview will be the main source for our story. Excerpts will introduce each chapter and will be used as flashbacks on George Gamow's incredible life.

Our journey along Gamow's *world line* also represents an opportunity to understand how the trajectories of physics combined with the dramatic events in the historical period between the late 1920s and the 1950s, and how much the global political situation of the time influenced its development.

●2

# Just Looking at the Sky

*Weiner: And when your father gave you the telescope, what did you plan to do with it? Did you know what to do?*

*Gamow: Just look at the sky.* [AIP]

## Up and Down the Potëmkin Stairs

Odessa, current Ukraine. When Georgij Antonovič Gamov (later trans-literated to George Gamow) is born, it is part of Novorossia, a region belonging to the Russian Empire under Tsar Nicholas II.

On March 4, 1904, Mrs. Alexandra gives birth to George by means of a complex C-section that must urgently be performed at home, on his father's desk and under the light of a kerosene lamp. The circumstances of his birth, as Gamow recalls them in his autobiography [GG70], in his father's studio and surrounded by bookshelves, were somehow warning signs of an absolutely unusual and extraordinary life to come.

In times when families were normally large, this family consisted in just three people, because George's difficult birth did not allow further pregnancies. The family photo that can be found at the beginning of the autobiography, shows little George with his parents: some details, such as their elegant clothing, the background and the decorations all around them indicate quite clearly that the Gamows were a wealthy family.

Indeed, his father Anton is the descendant of Russian Army officers, and a teacher of the Russian language and its literature in a boys' school in Odessa. His mother is the daughter of a Metropolitan, a high ecclesiastic rank of the Orthodox church, and she teaches history and geography in a girls' school in Odessa. An uncle on his mother's side was a distinguished chemist and George's only ancestor somehow involved in science.

George is raised in a family that constantly stimulates him. From a very early age he develops special talents, both in terms of learning speed and memory. He soon shows the desire to learn, to solve problems and to understand the mechanisms of things around him. He loves to make experiments, both with his microscope and his telescope, special gifts he received from his father in different moments of his childhood.

From his mother, he receives a classical education: she reads books to him every day, and his favorite author is Jules Verne.

Besides just understanding, George soon feels the need to test and demonstrate what he is learning. As he writes in his autobiography, he had already decided to become a scientist when a child, and on a very precise occasion, that also coincides with the end of his religious faith. One day, at the end of a function, he ran out of the church holding in his mouth the consecrated host he had just received. Once at home, he analyzed it with his microscope and compared it to a normal piece of bread dipped in wine. Verifying that there was no evidence of a difference between the two, he concluded that the idea of transubstantiation had failed. Hence, he gave up his religious faith.

Regarding Odessa, one's thoughts may immediately go to the *Battleship Potëmkin,* the famous movie directed by Sergei Eisenstein. In fact, one of its most dramatic scenes, the repression of the popular insurrection during the 1905 Revolution and the pram going down the steps unattended, was filmed on the Odessa Stairs.

George, obviously, could not have direct memories of those days, but he did remember very clearly the times of the other revolution, that of 1917. He shares many memories about those days in his autobiography: hard times, with violent fights between different ethnic Russian factions and against occupying troops of various nationalities.

The complex ongoing historical phase strongly affects everyday life. The situation is extremely uncertain and fundamental goods are difficult to find. A serious economic crisis has hit Russia, and the Gamows, even though originally a wealthy family, are severely affected by it too.

George's mother dies when he is just nine years old. Due to the war, his father loses his job as a teacher, and accepts to work in the same school as a janitor, to provide for his son's studies. George adapts very well to the difficult situation, even if he is just a child. This is an example.

> Potable water is lacking in Odessa, and its inhabitants must go down to the harbor every day to get water from the only public taps available, placed on big tanks that bring water to Odessa from the river Dnestr. George is in charge of supplying water at home, and every day he walks down and then back up the 200 steps that from Primorskij Bulvar lead down to the harbor, carrying heavy buckets filled with water.

This is the atmosphere in which George grows up. Despite all difficulties, he develops a very positive and optimistic character (*solution oriented*, as we would say nowadays). He is outgoing and relates very easily to other people, something that will turn out to be very helpful in his professional career.

In 1914 the First World War breaks out. George is 10 years old and attends school, but all of a sudden, studying is not easy. Classes cannot be given regularly, and George must study on his own; the many resources available at home are very helpful. He loves poetry, both Russian poets and foreigner authors, such as Shakespeare, Goethe and Schiller. Thanks to his outstanding memory skills, he will remember long passages of different poetic works even many years later. Something he really likes to show off, throughout his life.

As for mathematics, since childhood George shows extraordinary intuitive skills in problem solving. It is during high school that he first hears about the new Theory of Relativity: Einstein had published the Special Relativity in 1905, followed by General Relativity in 1915. Gamow is absolutely fascinated by this theory, a revolutionary idea that completely changes the interpretation of reality. Understanding such topics would have required a level of mathematics way beyond that of a high school student. Yet, young George is so motivated and passionate about relativity, that he starts studying differential calculus on his own, at a time when his peers at school are still studying algebra.

In the meantime, in 1917, the Russian Revolution has broken out, and the situation is getting even more chaotic: troops face each other in town, and bombings interrupt school lessons as windows crash. Classes must be suspended for long periods of time.

Gamow graduates from high school in 1921 and enrolls at the Faculty of Mathematics and Physics of the University of Novorossia, in Odessa. Lessons are very irregular, made difficult by recurring blackouts due to lack of fuel. In Odessa, mainly maths courses are offered, such as advanced algebra, multi-dimensional geometry and mathematical foundations for physics. He particularly appreciates one of his teachers, professor Schatunovski. Gamow had been impressed by an answer that professor Schatunovski had given to a student, who had noticed a mistake in a simple calculation on the blackboard. "Making correct calculations is not the task of a mathematician", professor Schatunovski replied furiously, "it's the task of an accountant!". Struck by this answer, Gamow will make it his own, and transform it into a fundamental principle for his professional career: whenever an idea is good, a trivial mistake is allowed.

The maths courses at the University of Novorossia are good overall, but the problem is that no physics courses are offered. There is only one physics teacher there, professor Kasterin, but he refuses to teach, as the available laboratories lack all the necessary equipment.

## Petrograd is Well Worth the Family Silver

So, after just one year of attending classes in Odessa, Gamow has no choice but to move to Petrograd, to continue his studies in a university that better suits his needs. The University of Petrograd (the new name of St. Petersburg after 1914) meets all his requirements. George's father Anton has a good friend there, professor Obolenski, a teacher of Meteorology; he might be a good referent. Moving to Petrograd is quite expensive and since the Gamows now have limited economic resources, Anton must sell the family silver.

Young Gamow thus moves to Petrograd in 1922. Upon leaving Odessa he has just one regret: he never declared his true feelings to Tania, professor Kasterin's daughter. In the future, Gamow will overcome his natural shyness towards girls.

The decision to move to Petrograd turned out to be crucial for Gamow. The scientific environment there was lively and innovative, mainly thanks to the contributions of Abram Ioffe and Paul Ehrenfest.

Born in Ukraine in 1880, Abram Ioffe had studied physics and engineering, first in St. Petersburg and then in Munich, where he obtained his PhD and got in contact with Wilhelm Röntgen, the discoverer of X-rays.

Paul Ehrenfest, same age as Ioffe, was Austrian. He had been trained in Wien, his hometown, at the school of Ludwig Boltzmann, the great scholar in thermodynamics and statistical mechanics. In 1901 he moved to Göttingen to continue his studies; there he met Tatyana Afanasyeva, a young Ukrainian-born mathematician, who became his wife in 1904. Together they moved to St. Petersburg three years later.

These two young scientists, Ioffe and Eherenfest, already knew each other, as they had met in Munich a couple of years earlier. In 1907, they meet again, this time in St. Petersburg. They are both expert and enthusiastic scientists, Ehrenfest a theoretician, while Ioffe an experimental physicist. They attract the attentions of many of the younger scientists in St. Petersburg, eager to discuss the incredible news that circulate among physicists at that time: above all, statistical mechanics, quantum mechanics and special relativity. And prudentially they never involve senior professors like Ivan Borgman and Orest Khvol'son, in their discussions.

However, despite their unquestionable abilities, Ehrenfest and Ioffe encounter many problems on their paths towards an academic career in St. Petersburg. Prejudices linked to his Jewish origin and his agnosticism play against Ehrenfest's career. A distinguished scientist, he runs for academic positions in several other European universities, and in 1912 he accepts the chair of Theoretical Physics (that Hendrik Lorentz has just left) at the University of Leiden in Holland. That's where Gamow will meet Ehrenfest the first time.

Ioffe somehow manages to access an academic position in St. Petersburgh, after some initial difficulties. As an assistant professor, he attracts to St. Petersburg many young and talented students. In order to increase the potential of the department, he purchases experimental equipment from abroad. Within a few years, Ioffe reaches the top of his career, when he is elected a permanent member of the Academy of Sciences of the USSR. His efforts in the reorganization of physics in Russia were enormous, considering the number of institutes and agencies he had given birth to, during those years.[1]

---

[1] A detailed description of the evolution of scientific institutions in Saint Petersburg between the 1910s and the 1920s can be found in reference [St].

Based on all this, the premises for Gamow to feel in the right place at the University of Petrograd were all there.

Upon his arrival at the new university, besides attending the lessons, George must find a job to support himself. A first job offer comes from his father's friend, professors Obolenski: he offers Gamow a position as an *observer* at the institute of meteorology in Petrograd. Working as an observer mainly consists in carrying out meteorological measurements three times a day, 20 minutes per shift. Gamow does this very carefully, but with very little interest. Although this job leaves him with a lot of free time, that he can dedicate to his own studies, the prospect of becoming a meteorologist, as Obolenski is suggesting, is definitely not an exciting one. His true desire, Gamow says, is to become a theoretical physicist.

So, within a relatively short time, he puts an end to this activity, and accepts a post as teacher of physics at the Field Artillery School of the Red Army. He must teach artillery officers the essential elements of physics in order to shoot cannon balls and hit targets with precision: trigonometry, trajectory calculations and so on.

> This experience is full of funny episodes. First of all, being a teacher for the Red Army implies being promoted to the rank of colonel. The reason is that salary must be adjusted to hierarchical position. The consequences of Gamow's promotion will turn out to be quite grotesque, as he tells in detail in his autobiography.
>
> One of the funniest events of that period takes place when Gamow must cover for an officer fallen ill, and has to calculate the precise trigonometric coordinates of a number of "targets" placed in the countryside surrounding the School of Artillery. Fake cardboard targets shaped as churches (a bit of anti-religious propaganda is always good, Gamow notices) had been set up for a shooting exercise. Having carefully made

all the measurements, Gamow hands out the coordinates of all targets to the commander in charge of the sighting operations. But just a second before the order to shoot is given, one of the supervisors notices something strange: there are seven targets on Gamow's list, while only six cardboard churches had been placed. Realizing what was about to happen, the supervisor screams "Stop! Position number 7 is not a target, it's a real church!" and avoids a tragedy in the very last moment. Gamow always loved to tell this anecdote, as it was a good opportunity to make fun of both Soviet institutions as well as of his own awkwardness, especially when it came to everyday situations.

Having held the rank of colonel of the Red Army, even though simply in order to receive a suitable salary, will end up as a rather relevant precedent in Gamow's life. When he is a professor at the George Washington University, already an American citizen, because of this he will be excluded from the Manhattan Project, despite his excellent competence in nuclear physics.

Meanwhile, Gamow attends the classes at the University of Petrograd, and complains about having to include two "political" exams in his curriculum, namely "History of World Revolution" and "Dialectic materialism".

In 1925, one year ahead the traditional schedule, he graduates, with the highest degree. He can now begin his research doctorate, but professor Dmitri Rogdestvenski, the director of the Physics institute, suggests him to wait one more year. In the meantime, in order to ensure him a salary, he offers him a one-year position at the State's Optical Institute. Gamow's tasks there are technical ones: he is in charge of selecting good quality optical components and of analyzing spectroscopic properties of gases.

After a year, the time comes when he can finally go back to his own studies, and Gamow chooses a research doctorate in experimental physics. It was a *strange* choice — Gamow says. And strange it was, since it was motivated only by the fact that experimentalists were allowed to have their own room at the institute, where they could hang their coats. Theoretical physicists did not have a designated room, and therefore had to leave their coats outside. Literally.

After making a number of disasters in the lab, however, Gamow finally realizes that experimental physics is not for him. Regardless of his coat, he will orient himself towards theoretical physics.

## The Three Musketeers

Luckily, Gamow is not the only student with an inclination for theoretical physics. In 1924, both Lev Davidovič Landau (Dau) and Dmitri Dmitrevič Ivanenko (Dimus) have reached Leningrad (the new name of Petrograd after 1924). Together with Gamow, whose nickname was Geo (to be pronounced as Joe), they establish a friendship that will last in time: they call themselves the *three Musketeers*. A couple of years later, another young and very talented Ukrainian physicist, Matvei Bronstein, will join the group — we will meet him again further on in this book. These young and brilliant scientists are especially interested in theoretical physics and are all engaged in studying and discussing the latest discoveries within quantum mechanics and relativity. But not only.

> The atmosphere among the students in Petrograd is very lively and stimulating and, as quite typical, also goliardic. Besides sharing the same scientific interests, the *three Musketeers* also like motorcycles, music and girls.

Many photographs of those years can be found in Gamow's autobiography as well as in other collections. They show boys and girls in funny poses, clearly having fun. In one of the photos, Gamow sits with his friend Zhenia on his lap, as Landau plays the cello. Gamow is wearing knickerbockers and long checkered socks, elements that reveal a taste for quite bizarre and unusual outfits. In another photograph, taken some years later during an excursion by boat on a Swiss lake, Gamow stands in short pants and long socks close to physicist Wolfgang Pauli, in a dark elegant suit. Despite his outfit, Pauli seems to be very interested in what his young and original colleague is telling him.[2]

At the University of Petrograd, students meet to discuss physics, or just hang out together, in two rooms at the Borgman Library (donated upon his death by before mentioned professor Ivan Borgman). The place is warm and the atmosphere is pleasant and relaxed.

Gamow, Landau and Ivanenko appear officially for the first time in the world of research, during the meetings of the Russian Association of Physicists. Abram Ioffe is the organizer of conferences, both in Leningrad and in Moscow, that gather scientists both from Russia and abroad. At the 1926 Moscow conference Gamow, Landau and Ivanenko have the chance to present some of their contributions.

That same year Gamow publishes, together with Ivanenko, his first work in an important international journal [GI]. The topic is very abstract: a five-dimensional space, the first four dimensions being represented by space-time, while the fifth is a function, typical of quantum physics. Gamow will never

[2] This one, as well as other interesting photographs of Gamow, can be found in Otto Stern Photograph Collection, http://bancroft.berkeley.edu/.

again engage in such formal subjects. He will rather focus his research on phenomenological aspects, motivated by the desire to understand the nature of physical phenomena and to describe them in the simplest possible way.

After the short and disappointing experience at professor Rogdestvenski's optical laboratory, Gamow realizes that he must follow his natural passion for relativity. Having studied mostly on his own, he now feels the need to get a more solid mathematical basis. He looks around and finds a course that seems to perfectly fit his requirements. It's professor Aleksandr Aleksandrovič Friedmann's course, whose title sounds like music to Gamow's ears: *Mathematical foundations of the theory of relativity*.

When Gamow signs up for the course, in 1924, Friedmann is considered as a real star at the University of Petrograd. Born in St. Petersburg in 1888 from a family of artists, he first studied mathematics and physics in his hometown. During the First World War, he had taken part in risky missions on military aircrafts, where his task was the application of effective bombing techniques, based on ballistic properties. Starting from 1918, he had been a professor first at the University of Perm, in eastern European Russia. Then, a couple of years later, he became professor of mathematics and physics, back at the University of St. Petersburg.

Here, Friedmann dedicates himself to the study of general relativity and to its applications to cosmology. He feels he must make up for the time he has lost, partly due to the war and partly due to the delay with which foreign scientific journals reach Russia.

His attention is drawn to Albert Einstein's article of 1917. In it, he had formulated a space-time model of the Universe, based on his relativistic equations. Einstein's idea is that the cosmos has properties of homogeneity (all points are equivalent) and isotropy (all directions are equivalent), but also that the relative distance between two points in space does not vary with time. For this last condition, Einstein was strongly influenced by the idea, very common at that time, that the Universe is *static*.

Friedmann, however, starting from Einstein's relativistic equations, works on their general solutions and derives two possible models. The universe either expands indefinitely or, after a phase of expansion, starts contracting. In this case, it could result in a periodic universe, where phases of expansion and contraction alternate. Friedmann's dynamic models of the universe are the opposite of Einstein's static one. He publishes his conclusions on the *Zeitschrift für Physik* in June 1922.

Founded in 1920, the *Zeitschrift für Physik* soon became the most renowned physics journal in Europe, and helped defining German as a language commonly used within the scientific community. It also was the reference journal for the scientific community in Russia, that was rather isolated from the rest of the world. The situation changed only after the Second World War when, as a consequence of Germany's defeat, the international scientific world chose English as universal language.

When Einstein reads Friedmann's article, he immediately prepares a reply — his comments are published in the September 1922 issue of the same journal. Einstein states that there is a mistake in Friedmann's calculations and that the non-static solution is not compatible with general relativity. This could have represented the tombstone for Friedmann's ideas, but he is a brave and very self-confident guy. Careful and respectful in front of a personality such as Einstein's, Friedmann does not send his own counter-arguments to the journal, but addresses a letter directly to Einstein. In it, he explains in detail the reasons why he is convinced that his calculations are correct. And adds: if my explanation persuades you, please send a correction note to the journal. Friedmann also asks him to meet in person, but is never given a chance to.

Einstein will read Friedmann's letter many months later, when returning to Berlin after a long journey abroad. In Leiden, Einstein had come across professor Yuri Krutkov, a colleague of Friedmann, who had further illustrated the reasons why those derivations were correct.

Einstein is finally convinced. In May 1923 he sends a note to the *Zeitschrift für Physik*, in which he admits having made a mistake, defining Friedmann's results as *enlightening*.

Despite all this, Einstein didn't give up his static model. He will reject it only many years later, when astronomical observations clearly indicate that distances between galaxies increase as a function of time.

Besides relativity, Friedmann was also engaged in other fields of research; among these, modeling the dynamics of the atmosphere. In order to do this, some physical parameters had to be detected by using aerostatic balloons at very high altitudes. Friedmann himself flies on one of these balloons, together with aviator Fedosenko, setting an altitude record of 7,400 meters.

Gamow's decision to attend Friedmann's class was obviously motivated by his personal interest for relativity and cosmology and by the possibility of having him as a supervisor for his PhD. Besides, he certainly was also curious to meet such a great scientist in person. But fate gets in the way. In September 1929 Friedmann suddenly dies. Some say due to typhoid fever, others to a severe pneumonia, contracted during a balloon ride. For Gamow, Friedmann's death implies postponing his *rendez-vous* with relativistic cosmology until the second half of the 1940s, when he will come up with one of his most extraordinary scientific inventions.

Given the sudden and unexpected change of plans, Gamow's prospects are not so enthusiastic anymore. Professor Krutkov has replaced Friedmann as his PhD supervisor and suggests for his thesis a quite boring mathematical-physical topic: *Adiabatic invariance of a quantized pendulum with finite amplitudes*. Certainly not much for Gamow — the title itself must have sounded terrible to him.

Gamow finds a way out thanks to old professor Orest Khvolson, now retired: having noticed young George's talent as a student, realizing the difficult situation he recommends him for a scholarship to study abroad. And he also indicates the destination: Göttingen. Professor Khvolson's request is supported by other professors at the University too, among which Yuri Krutkov, and Gamow is awarded the scholarship.

At the beginning of summer 1928, he is ready to leave for Germany.

●3

# The Next Day the Paper Was Ready

*Gamow: Well, you see I can tell you how I came to this idea about the potential barrier [...]. I went to Göttingen for the summer school and I didn't go to any lectures. At this time everybody was applying wave mechanics to atoms and molecules and getting more and more complicated, and I hate such things [...]. Then I decided that I wanted to learn something about the nucleus. In the library in Göttingen I found the Philosophical Magazine, with the article of Rutherford in which he was shooting the fast alpha particles [...] and Rutherford had a theory in this paper. Well, at this point when I read this, I thought, "My gosh, it isn't." It is exponential solution of the equation of Schrödinger. And the next day the paper was ready.*

*W: Really? You mean that this thing seemed to you ridiculous, in a sense, did it? [...]*

*G: [repeated several times during preceding question] Yes. [AIP]*

## The First Sanctuary: Göttingen

This was the opportunity he had been waiting for. He finally had the chance to measure his potential in an international research environment. At that time, the Institute of Theoretical Physics at Göttingen was considered by young physicists as one of the most prestigious universities. It was directed by Max Born, one of the greatest theoretical physicists of the time. Remarkably, less than three years earlier, Born had laid down the foundations of one of the pillars of quantum mechanics (matrix mechanics). He had been prompted to do this, by the extraordinary intuition of a young theorist, namely Werner Heisenberg, and together they further developed the theory with one of Born's students, Pascual Jordan. Born had then moved further to the development of a statistical interpretation of quantum physics.

In June 1928, on his departure for Göttingen, Gamow shares his enthusiasm for the upcoming journey with his friends. A cheerful group of boys and girls accompany him at the harbor, as he boards the steamboat that will take him across the Baltic Sea, from Leningrad to the German port of Swinemünde. From there, he will then reach Göttingen by train, the following day.

Gamow provides a detailed description of his first evening in the German city — he obviously has a nice memory of that night, even many years later. Having reached Göttingen late in the afternoon, he leaves his luggage at the train station and meets his friend Yakov Frenkel, a professor from Leningrad who is there for the same summer school. Since Frenkel had arrived in Göttingen some days earlier, he has already made friends with some of the other participants. That night, they are invited by Max Born to a party in a local pub. Beer, food specialties, dances and games: enough to convince Gamow to join the company. Hours pass and it gets really late. Gamow suddenly realizes he must find a place to spend what remains of the night. He luckily finds a room at the local inn, and does not even mind that his luggage is still at the train station. Not a bad way to start!

The scientific environment that welcomes Gamow in Göttingen is extremely lively. The people at the Institute for Theoretical Physics, as well as almost the entire town, are enthusiastic about the new theory of quantum mechanics. Groups of physicists sit and discuss it even in cafés. But something bothers Gamow: quantum mechanics is already becoming a topic for formal discussions and philosophical speculations — something he is definitely not interested in.

It is quite curious to notice that Enrico Fermi, who also had been a scholar in Göttingen just a few years earlier, never got really acquainted with the local environment either. Everyone seemed to be engaged in philosophical debates about the new physics. Neither Fermi nor Gamow showed any interest for the formal aspects of physics, but rather preferred a markedly phenomenological approach.

On the side of phenomenology, at Gamow's times the research carried out at the Institute in Göttingen focuses on the applications of quantum mechanics to atoms and molecules. But Gamow considers this field of research way too crowded, everyone seems to be involved with this in Göttingen. He therefore decides to look for something different to work on, on his own, something new and more interesting. As physicists in search of inspiration used to do in the past century, Gamow goes through the latest issues of the most important scientific journals he can find at the Institute's library, in search for some new experimental result to be investigated (nowadays, everyone can easily access all new articles online, on updated international archives).

On a day of June 1928, Gamow comes across the latest volume of *Philosophical Magazine,* and his attention is drawn to an article written by Rutherford.

Obviously, the fact that Gamow starts reading an article by Rutherford is not a coincidence. Ernest Rutherford is a celebrity in the field of experimental physics and Gamow is very familiar with the research he has been working on. In the first years of the century, in fact, Rutherford had formulated, together with chemist Frederick Soddy, the physical laws of $\alpha$ and $\beta$ *radiations* emitted by radioactive substances. At the time of his research with Soddy, Rutherford, who was a native of New Zealand, was in Montreal, Canada, as a professor of Experimental Physics at the McGill University, after having received a fellowship at the Cavendish Laboratory in Cambridge.

After some years of experimental research, Rutherford found out that $\alpha$ radiation consists of emitted particles — $\alpha$ particles — with a mass about four times that of the hydrogen atom, and with two units of positive electric charge (that is, an $\alpha$ particle is the nucleus of a helium atom).

Since then, Rutherford systematically used the $\alpha$ particles emitted by radioactive elements as bullets to be sent against a target, consisting of a thin foil of various chemical materials. This experimental procedure turned out to be a very efficient way to explore the atomic and nuclear structures of matter, and allowed Rutherford to come up with two major results.

In 1911, together with Johannes Geiger and Ernest Marsden, he set up an experiment where he measured by how much the alpha particles impinging on the target were deflected, as a consequence of their collision. From the recorded results, he inferred that the positive electric charge of the atom is concentrated in a tiny central region, whose size, in diameter, is four to five orders of magnitude smaller than the atomic size: the idea of *atomic nucleus* was born.

Rutherford obtained a second important result some years later, in 1919: he showed that by bombarding nuclei with $\alpha$ particles, it is possible to induce nuclear disintegration. In this process, a positively charged hydrogen atom (or, more precisely, a hydrogen nucleus) is emitted — this is the particle that Rutherford, at a meeting of the British Association in August 1920, called a *proton*.

All this gave birth to Rutherford's model of the atom: a neutral atom consists of Z electrons orbiting around a tiny central core, made of A protons and A-Z *nuclear electrons*. According to this model, an $\alpha$ particle is a combination of four protons and two *nuclear electrons.* Rutherford intuitively also understood that a proton-electron system inside the nucleus can simply be considered as *one* particle or, to use Rutherford's words, a *neutron.* One of his colleagues, James Chadwick, will prove him right experimentally in 1932. We will return to this further on.

But let's go back to the library in Göttingen where Gamow is about to start reading Rutherford's latest article: *Structure of the Radioactive Atom and the Origin of the $\alpha$-Rays* [Rut]. This article represented, in the author's intention, the theoretical explanation of processes induced by hitting nuclei with $\alpha$ particles and of the mechanism of natural radioactivity. In this publication Rutherford draws on his 30 years of experience in this particular branch of physics.

## Nuclear Physics Becomes Quantum

Gamow reads the article carefully and is stunned by the complicated arguments that the author uses to explain how an $\alpha$ particle, initially confined within a heavy nucleus (such as uranium), can freely escape. Why does Rutherford imagine such complicated and unnatural mechanisms — Gamow argues — when the explanation is so simple in terms of the new quantum mechanics?

We are here on the border between the old and the new generation of physicists. There are 35 years of age difference between Rutherford and Gamow and, in terms of scientific development, a huge leap between classical physics and the innovative principles of quantum physics. Gamow is 24 years old and has just completed a full immersion in the new physics in Leningrad, together with his *Musketeer* friends; he immediately understands that $\alpha$ decay is a direct manifestation of the so-called quantum mechanical *tunnel effect*. This quantum effect had just been investigated by two other young physicists, Lothar Nordheim and J. Robert Oppenheimer, in order to explain the conduction properties of electrons in metals and the emission of electrons from cold metallic surfaces, by means of intense electric fields.

Gamow ventures into the application of quantum mechanics within the brand-new theoretical field of nuclear physics; by so doing, he lays the foundations of a theory that will be further developed in the upcoming decades. As to $\alpha$ radiation, according to Gamow the process can be described as follows. The $\alpha$ particle is bound by attraction forces inside the nucleus — to use Gamow's image, it is as if it was confined within a fortress surrounded by tall walls. It can move back and forth but, unless its energy is big enough to allow it to pass above the walls, it will remain confined inside the fortress. However, quantum mechanics offers the $\alpha$ particle a possibility to escape. According to this theory, in fact, the particle has a probability, maybe tiny, but not nil, to *leak through* the wall — that is, it can escape by *tunnel effect*. In fact, once it has penetrated the barrier, that is when the $\alpha$ particle has come out of the nucleus, its subsequent escape is eased by the Coulomb repulsion between the positive charges of both the remaining nucleus and the $\alpha$ particle itself.

Experiments on $\alpha$ radiations measure both the half-life of the radiative process (that is, the time needed for the number of atoms present in a sample of pure radioactive material to halve via decay) and the energy with which the $\alpha$ particle is emitted. Both these values vary — also sizably — from a kind of nucleus to another, but are correlated. According to an *empirical* formula obtained by Hans Geiger and John Nuttall in 1911, the half-life is greater the smaller the energy of the emitted $\alpha$ particle is. For example: the half-life of uranium-238[1] is 4.5 billion years and the energy of the $\alpha$ particle is 4.1 MeV, while polonium-214 has a half-life of 0.0002 seconds and emits $\alpha$ particles with an energy of 7.7 MeV (1 MeV = 1 million electron-volts; approximately, 1 MeV is the energy equivalent to twice the electron rest-mass, i.e. 1 MeV $\approx$ 2 $m_e$ $c^2$).

Gamow can demonstrate that his approach to nuclear $\alpha$ decays is correct, as he manages to derive the correct quantitative correlation between the half-life and the energy of the $\alpha$ particle, as represented in the formula that Geiger and Nuttall had empirically obtained.

He shares his result with the people at the Institute in Göttingen: he discusses it with Eugene Wigner, a young Hungarian physicist, at that time assistant to the great mathematician David Hilbert, as well as with his colleagues Friedrich Houtermans and Léon Rosenfeld — and, most remarkably, with Max Born.

Born immediately asks Gamow to give a seminar on this subject. Many years later, Rosenfeld describes in an interview the strong impact Gamow's presentation had had on the audience. Born's first reaction was: "That's a very great discovery. Mind you, this is a very general method for quite a lot of phenomena." [AIP1]

However, the formal derivation of the result still left Born somewhat puzzled. He thought it could be done better, and put himself at work the following year — he derived Gamow's same result in his own way, *à la Born*.

---

[1] In the present notation, the number following the name of a chemical element is the mass number A, i.e. the number of protons plus the number of neutrons forming the atomic nucleus. It is worth recalling that a chemical element, identified by the *atomic number* Z, can occur in nature in various nuclear compositions, called isotopes, characterized by different values of the *mass number* A.

Gamow seems quite surprised by the hustle his idea has caused in Göttingen. If he was truly surprised or just pretending, is something we will never know. According to Rosenfeld, Gamow had probably already elaborated upon the basic idea when he was still in Russia: "He came directly from Russia and had practically the whole thing when he came here. It may be that he got the final solution or worked out the consequences in Göttingen. That is very probable. But he had certainly the main idea; he had it already when he came." [AIP1]

Whatever the truth is, this is a good example that helps us understand Gamow's peculiar intelligence and original approach to things and people. Far from being a naive young boy who had come to Göttingen, the "Mecca" of physics, in search for new ideas, he had already worked on his own ideas and was here in order to find a renowned audience to share them with.

Gamow writes his article about $\alpha$ decays and, considering the level of accuracy with which he wants to describe the phenomenon, he does it very fast. Reading that article today might make us smile. Many of the calculations presented therein are those that physics' students nowadays learn during their third year of studies. At that time, however, they represented the very first application of quantum physics to the newly born nuclear physics.

That Gamow had worked on his discovery independently is quite obvious from what he writes at the end of the article: he does not mention having discussed his idea with Born, but simply thanks him for the *hospitality received at the Institute.* Born's liberal policy granted all visiting physicists the complete freedom to publish their papers under the aegis of the Institute. The fact that Gamow *thanks for the hospitality,* anyway, implicitly meant that Born had at least read the article beforehand and was informed about its content.

On July 29, 1928 Gamow's article for *Zeitschrift für Physik* is ready [GG28]. This date is worth being remembered, because on the following day, July 30, 1928, Ronald Gurney and Edward Condon (both researchers at the Palmer Physical Laboratory of the Princeton University) send to *Nature* a scientific note, presenting a similar idea [GC]. Because of the different press times of the two journals, Gurney and Condon's article is published a month before Gamow's.

> The *race* to publish scientific articles in specialized journals as fast as possible has always been typical of the spirit of competition that characterizes the world of research, especially in physics.
>
> Nowadays, the web witnesses the exact moment in which a result is made public among the scientific community. A scientific paper then becomes more or less visible and appreciated, depending on the prestige of the journal that accepts to publish it, on the basis of careful evaluations carried out by its referees.
>
> At Gamow's time, the two leading scientific journals in Europe were the before mentioned German *Zeitschrift für Physik* and the British *Nature*. The latter was normally used for shorter notes, that introduced to more detailed articles to appear in *Zeitschrift für Physik* or in the American journal *Physical Review*.

Gamow details his explanation of $\alpha$ decay in terms of tunnel effect in a series of articles he writes between September and November 1928. The September paper is signed together by Gamow and Houtermans.

Friedrich Houtermans, a brilliant young physicist who also was working in Göttingen, had convinced Gamow that his theory of $\alpha$ decays needed to be thoroughly formalized. Therefore, they quickly wrote an article together for *Zeitschrift für Physik* [GH]. Gamow was probably not sure that this further investigation could raise any particular interest. Most likely, he had agreed to write the article just to please Houtermans, who was an experimental physicist now intrigued by theoretical subjects. Indeed, it sounds funny that many years later, when he runs into this article during the interview with Charles Weiner, Gamow comments it as follows: "And this is Gamow and Houtermans, and this is just tedious, more details, calculations of the things, just follow up" [AIP].

> Very much has been written about Friedrich Houtermans and his incredible life. We here simply want to notice that Gamow and Houtermans were very close friends, and shared a similar genius and recklessness. Gamow liked his friend Fritz's habit to sit and work at the café, at a table covered with sheets of paper and calculations, among a dozen empty coffee cups. The waiter used to collect all the empty cups at the end of the day, he counted them and brought Fritz the bill accordingly. To him, Gamow's intuition as to the capability of an $\alpha$ particle to penetrate the nuclear barrier sounded somewhat extraordinary.

Gurney and Condon also write a second and more detailed paper, in which they further elaborate upon the results of their research. However, they make a big mistake: they do not consider the possibility that tunnel effect might be relevant in the case of the *capture* of an $\alpha$ particle by a nucleus — that is, in the *backward* process compared to the emission of an $\alpha$ particle. Gamow demonstrates that tunnel effect is relevant also in this case, and by so doing opens up a completely new field of research, that we will discuss further on in our book.

Gamow's theory of $\alpha$ decay will arouse enormous interest, inspiring a great number of articles by many different authors. As Gamow recalls in his interview: "I remember Pauli, who was there at this time, whenever new papers [appeared] — there had been half a dozen of them — he used to say, *"Es Gamow't wieder"* [It's Gamow again], like *"Es regnet wieder."* [It's raining again]."

In September 1928, once their paper is ready, Houtermans and Gamow prepare to leave Göttingen. Houtermans is heading to Berlin, whereas Gamow back to Leningrad. But not immediately, since he still has some (not much) money left over, he wants to stop off in Copenhagen, and get in touch with another charismatic physicist: Niels Bohr.

Gamow's description of his lucky and apparently fortuitous encounter with Bohr is typical of the author's creative storytelling. According to his version of the story, Gamow shows up unexpectedly at Bohr's office in Copenhagen. After insisting with his secretary, he gets an appointment and has a chance to talk to the great master. Bohr hasn't come across Gamow's article on $\alpha$ decays yet, and asks him what he is currently working on. Evidently impressed by the subjects this young physicist is explaining, and having been informed that this young Russian guy is probably broke, Bohr offers him a Carlsberg scholarship, in connection with the Royal Danish Academy of Sciences, for the upcoming academic year 1928–1929. An offer that Gamow obviously accepts straightaway, enthusiastically.

It is a nice story, and if it is not true, it is well made up. There are good reasons to believe that the events had probably had a different turn. But Gamow's reconstruction of the story is so amusing that it almost feels tactless to remember that he had contacted Bohr already in the previous month of July, to explore the possibility of spending a period of time in Copenhagen. And that as a support to his inquiry, he had sent a presentation letter signed by Abram Ioffe.

# If I Had Been Smarter

*Gamow: An equation connecting the energy of alpha-particles with the surface tension of the imaginary "water-drop". This is where I could have predicted fission, if I'd been smarter.*

*Weiner: This was the time when you had stopped off to see Ehrenfest on the way to England [...]. What was the nature of the discussion with Ehrenfest?*

*Gamow: Nothing special. Well, essentially the droplet, the droplet model. I remember he told me that I should put it all down. And I didn't, so when Bohr and Wheeler formulated it, I was not quoted then.*

## From Copenhagen to Cambridge, Stopping Off in Leiden

Gamow immediately feels at ease in Copenhagen. Probably the mere fact that his scholarship is named after the famous Danish Carlsberg beer, is to him a source of great satisfaction. The atmosphere at the Institute directed by Niels Bohr, another "sanctuary" for physicists, is very friendly: Bohr likes to talk with his colleagues and students and has very informal relations also with the younger ones. Since Gamow doesn't hold a particular position there, he has no fixed working hours to respect. At the Institute, people are allowed to arrive late and stay as long as they want, switching between physics discussions and table tennis matches.

Bohr is very active and supportive of young people. He immediately realizes how brilliant this young physicist is. He has just arrived in Copenhagen via Göttingen, and deserves to be launched on the international scene. Considering Gamow's brilliant interpretation of Rutherford's results, Bohr suggests him to visit the Cavendish Laboratory at Cambridge University and get a chance to personally meet the *Crocodile* (Rutherford's nickname, coined by one of his students to underline his strength and determination).

Bohr knows from his personal experience that Rutherford does not particularly like theoreticians, and therefore writes a presentation letter of the young physicist to the *Crocodile*, to prevent his possible negative reaction. After all, Gamow has criticized Rutherford's theory of $\alpha$ decay.

Gamow's visit to Cambridge is quickly arranged for January–February 1929 and towards the end of 1928, Gamow leaves Copenhagen, heading to the Cavendish Laboratory. On his way, he stops off in Leiden, in Holland, as he wants to talk with Paul Ehrenfest (whom he had met in Leningrad some time before) about a peculiar model of the atomic nucleus he has been working on.

> Gamow's idea is the following. An atomic nucleus can be considered as a set of particles that mutually attract one another. Deep inside the nucleus, each particle is attracted by all surrounding particles, but the ones on the surface are only subject to an attraction force from its center. The result is that the surface of the nucleus has a role of containment, something similar to the surface tension of a liquid drop. Gamow thus applies to the nucleus the idea of a surface tension: according to this description, a nucleus is stable when its internal pressure (which is due to the kinetic energy of its various components, that is, to their movement), added to the Coulomb repulsive forces between components with the same charge, exactly balances the surface tension.
>
> Expressed this way, this *liquid drop nuclear model* might sound rather trivial. But it is not, since it allows Gamow to make

an evaluation of the binding energy with which a single nuclear component is tied up inside the nucleus. This binding energy is the energy that must be provided to the nucleus in order to extract one of its components.

Gamow calculated values that fit quite well the binding energies that had been measured experimentally by the British physicist Francis Aston, by means of a special instrument — a mass spectrometer — at the Cavendish Laboratory. Aston is one of Rutherford's senior colleagues — the *masters*, as they are called in Cambridge [St]. Besides being a physicist, he is also a great sportsman, a brilliant competitor in many different sports such as bike, tennis, golf, swimming, alpine sports. Successfully managing to combine science and sport, Aston won the Nobel Prize for Chemistry in 1922. But this is a different story.

Ehrenfest likes Gamow's idea and encourages him to further elaborate upon it and to write a paper about it. This stop in Leiden on his way to Cambridge, in order to double check his nuclear model, is a clear sign of Gamow's desire to reach the Cavendish Laboratory with something new to show to Rutherford. Clearly, he also got there with his own theory of $\alpha$ decay in his briefcase.

Gamow's strategy is definitely a winning one. Rutherford realizes quickly that this boy is smart. He especially likes the way this young theoretical physicist explains things: he makes physical principles clear, with very little use of mathematical frills. The two definitely hit it off.

## A Great Show at the Royal Society in London

At the time Gamow visits Cambridge, Rutherford is at the top of his scientific career. Since 1925, he is the president of the renowned Royal Society, whose motto is "Nullius in verba" — meaning that in science a principle of authority is unacceptable and every statement has to be supported by experimental evidence.

And it is precisely at the Royal Society that Rutherford decides to organize, on February 7, 1929, a general discussion on the following subject: *Structure of Atomic Nuclei.*

Rutherford invites his young Russian guest to join the discussion, to which some of his most important colleagues at the Cavendish Laboratory will partake. For Gamow, this invitation represents the official crowning in front of the *élite* of the British scientific community.

In his introduction, Rutherford describes the evolution of nuclear physics over the past fifteen years, that is, since the first discussions on the composition of the atom, that had taken place at the same Royal Society [Rut29]. He describes all research developments at his laboratory, with particular focus on the systematic studies on isotopes. Rutherford highlights Francis Aston's works, aimed at measuring nuclear binding energies.

At this point he introduces Gamow. After some words of welcoming, he gives him credit for the description of the basic properties of the forces that act among nuclear constituents: an attractive short-range force, that competes with a repulsive Coulomb force. He also reminds the audience of Gamow's crucial contribution to the quantum interpretation of nuclear processes involving $\alpha$ particles. Rutherford's general presentation is then followed by the speeches by various collaborators. Finally, it is Gamow's turn to talk and he starts explaining his new liquid drop nuclear model. His explanation is straightforward. He illustrates the physical principles of his model and provides a short mathematical derivation, underlining how his model fits quite well Aston's experimental results.

## A Missed Opportunity for Gamow

Gamow will further detail his *nuclear drop model* some months later, and will write a report, that Rutherford himself will forward to the Royal Society: *"Mass Defect Curve and Nuclear Constitution"* [GG30]. Surprisingly,

Gamow never fully acknowledged the real value of his own results. Instead of following Ehrenfest's advice and promote his idea by writing a paper for some renowned scientific journal, he drops the subject and moves on to something different.

Gamow will realize the mistake he made only many years later, expressing all his regret to Charles Weiner during the interview.

> The liquid drop model became fundamentally important in 1932, when the neutron was discovered, and from then onwards the nucleus could be simply described in terms of protons and neutrons, with no need for hypothetical *nuclear electrons*.
>
> Gamow's formula, that related the binding energy to the mass number, was then implemented by other famous physicists such as Werner Heisenberg in 1933 and by his student Carl von Weizsäcker in 1935.
>
> The ongoing experimental research provided complete information as to the binding energy, as we know it today. The energy by which a single component (proton or neutron) is bound to the nucleus is measured around 7–9 MeV. The curve that describes the binding energy (as a function of the mass number) shows low values corresponding to lighter elements; it then increases and reaches a peak around the nucleus of iron, and then decreases again for heavier elements. This is just the average trend: significative deviations occur for specific elements such as helium-4, in which two protons and two neutrons are bound very strongly. We will see further on how this explains the natural abundance of helium-4.

The funny thing is that, although Gamow described the atomic nucleus as a liquid drop, he never got the intuition that the nucleus could, if stressed by internal or external perturbations, eventually distort and start a process of fission. At that time, neither Gamow nor any other physicist thought of this possibility.

————————— 5 —————————

# The Heaven's Glorious Sun

*Gamow: Of course, Houtermans and Atkinson had a paper much earlier; they did it right after my paper on Zertrümmerung, still in 1928.* [AIP]

*[It came about in this way]. We decided that the best place to discuss the problem of energy release in stars would have been some sky resort in the Alps. Then, in early spring I took a train, picked up Fritz and Robert in Berlin, and all together we proceeded to a small town in the Austrian Alps.*[1]

## A Letter from His Friend Fritz

Upon his return to Copenhagen, Gamow receives a letter from Berlin: his friend Friedrich (Fritz) Houtermans informs him about his recent encounter with Robert Atkinson, a brilliant young physicist that he had met in Göttingen, as a doctoral student collaborating at an experimental research. Despite being an experimental physicist, Atkinson was thrilled by theoretical aspects and most of all by astronomy. So much that he had managed to study Arthur Eddington's *The Internal Constitution of the Stars* [Edd], on his own.

> Eddington's work had been published in 1926 and had immediately become a fundamental reference book for theoretical astrophysics. Simply when reading the preface, one is impressed by the author's far-reaching perspective, especially

---

[1] Excerpt from George Gamow's autobiography [GG70].

when he clarifies the aim of his work — to study the physical processes occurring "in the deep interior of the stars" in order to explain "the external phenomena accessible to observations". Eddington there also ironically points out how different his vision is compared to that of Shakespeare, who had written:

The heaven's glorious sun
That will not be deep-searched with saucy looks;

and expresses his hope that the above sentence doesn't need to be finished according to the quote from *Love's Labour's Lost*

Small have continual plodders ever won
Save base authority from others' books.

With great insight, Eddington clearly is the precursor of a new and exciting field of research, in which astrophysics merges with nuclear and particle physics — "The scope of the inquiry has grown so that it now involves much of the recently won knowledge of atoms and radiation, and makes evident the ties which unite pure physics with astrophysics."

Eddington also underlines how fast physics and astronomy have developed in recent times, and to further stress this point, he mentions he had to be very careful when listing all the most recent scientific results in his treatise.

A full chapter of Eddington's book is devoted to the source of stellar energy. Here, he goes straight to a very concrete point: how old is the sun? This issue had been a subject of discussion since mid-1800, and not only among physicists: the long-lasting disputations on the age of the sun had involved naturalists and geologists on one side and Lord William Thomson, first Baron Kelvin, on the other.

Lord Kelvin, a prominent figure of 19th century physics, had estimated the age of the sun in some tens of millions of years.

His evaluations were based on German physiologist Hermann von Helmholtz's line of reasoning: the emission of energy from the sun is due to the transformation of the star's gravitational energy into heat, during its contraction.

However, since the solar activity deeply influences both the biologic evolution and the geologic changes on our planet, observations in these fields lead to estimate the age of the sun in hundreds (not tens) of millions of years. Due to this, Lord Kelvin's theory was strongly criticized by both geologists and naturalists.

Among the scientists mostly disappointed by Lord Kelvin's estimates, was Charles Darwin, the great biologist and naturalist. According to Darwin's theory of the evolution of natural and vegetal species, a solar age of a few tens of millions of years would have been absolutely inconsistent. Darwin was very determined in proposing his evolution theory, and was not discouraged by Kelvin's evaluations. However, he did not want to openly contradict Lord Kelvin, who after all held the chair of *Natural Philosophy* at the University of Glasgow. Thus, he found a sneaky way out, by simply avoiding any reference to time estimations in the last editions of his famous treatise, *The origins of species by means of natural selection.*

Eddington's theory obviously doesn't take into account Lord Kelvin's argument, nor the idea that the sun acquires energy from outside — through meteorites or capture of radiation that propagates in space, as claimed by some. Eddington is convinced that there has to be an internal source of energy in the sun, which would both explain the entity of its energy emission, and make sure its inner temperature is kept high enough to prevent it from collapsing.

At this point, Eddington has a fundamental insight: solar energy is produced in *subatomic processes* occurring in its inner parts, based on the equivalence between mass and energy, as implied by the theory of relativity. He considers two of these processes as the most likely ones: the annihilation of protons and electrons and the formation of complex chemical elements from the merging of simple ones (that is, nuclear fusion). For this second hypothesis, he clearly mentions the fusion of four hydrogen atoms into one helium atom, as a reaction that can release energy, based on Aston's measurements of the nuclear binding energy. By so doing, Eddington already indicates one of the fundamental reactions that are responsible for nuclear combustion in stars.

In his letter to Gamow, Houtermans illustrates the theory about energy production in the sun that he and his friend Atkinson have worked out together, combining their individual expertise. Atkinson has learnt from Eddington that nuclear reactions can be sources of solar energy, whereas Houtermans has learnt from Gamow that a charged particle can enter into the atomic nucleus by penetrating — quantum mechanically — the repulsive Coulomb barrier. Combining this info, the two argue that in the inner parts of stars (where hydrogen is abundant) protons can trigger nuclear transformations that release a huge quantity of energy, as predicted by Eddington. They now have to quantitatively evaluate the expected effects.

## On the Austrian Alps Talking about the Sun

In this, Atkinson and Houtermans need a theoretical physicist to guide them in their calculations. They address a request for help to Gamow, who accepts with great pleasure. But there is a problem: they are in Berlin, whereas Gamow is in Cambridge — so why not meet in some ski resort on the Austrian alps, to talk about physics and go skiing together? Many years later, Gamow remembers how happy he was that there he did not need to dedicate too much time to physics, since "they were almost ready with their calculations, and the discussion did not impose on our skiing time" [GG70].

Houtermans and Atkinson were the first to suggest a concrete way to achieve thermonuclear processes in the stars [AH]. Although the idea was brilliant, at that time their quantitative evaluations were affected by a number of uncertainties in the theory.

In his autobiography, Gamow sounds very amused when he remembers how things really went during the scientific discussions he had with Atkinson and Houtermans on the Austrian alps. The reason is that he had made two major mistakes in his theoretical consultancy, and quite big ones. He had overestimated a nuclear quantity by a factor of about ten thousand and, at the same time, underestimated another quantity by about the same factor. Since in the final calculation these two values multiply, the (huge) mistakes miraculously cancelled out! Gamow comments the episode stating that "this mix-up probably represents one of the most striking examples in the history of science of a case in which rapid advance can suffer from the pitfalls of nonconsolidated by-passed ground" [GG70].

As a matter of fact, it took almost ten years before the understanding of nuclear physics and astrophysics allowed to figure out realistic calculations.

When Charles Critchfield and Hans Bethe took up this topic in 1938, once again Gamow was involved. We will come to this further on.

In April 1929, considering the fact that many of the physicists who had spent time at his Institute were planning to go to Copenhagen for Easter, Niels Bohr decides to take the opportunity to organize a big meeting over the holidays. Physicists gather in the spirit of a family reunion, in a very informal and friendly atmosphere. Léon Rosenfeld reports that Gamow, as expected, became the star of the meeting, with his funny jokes and extravagant creativity.

This conference was a great success and induced Bohr to organize similar events in the coming years.

Since the Carlsberg scholarship is about to come to an end by early summer, it is now time for Gamow to start planning the coming years. Upon both Bohr's and Rutherford's suggestions, he applies for a Rockefeller scholarship, to spend one year in Cambridge, starting from fall 1929. In the meantime, he will go back to Russia to spend the summer, as he has spared enough money as a student at the University of Leningrad.

Upon his return to Russia, Gamow is welcomed very warmly. Newspapers celebrate him as a distinguished Russian citizen, who has honored himself and his own country, with his notable scientific merits. After a stop in Leningrad, he goes back to Odessa to meet his old father. And then leaves for Crimea, a seaside resort near Yalta, to spend a two-week relaxation. A well-deserved rest for a warrior.

Gamow is assigned the Rockefeller scholarship. Having obtained the passport from the *Narkompros*, (the People's Commissariat for Education, i.e. the organism controlling all activities connected to instruction, research and culture in the Soviet Union), at the end of September 1929 he leaves for Cambridge, to spend the coming year at the Cavendish Laboratory.

# Back to Cambridge and Copenhagen

*Gamow: Rutherford asked me to calculate what is the chance of proton penetration. And I told him, what energy the proton needed [...]. It turned out one MeV is enough.*

*Weiner: Now he asked you on the spot, while you were there, you mean?*

*Gamow: Well, this was just during the period later on ...*

*Weiner: When you spent considerable time there?*

*Gamow: I think this was when I was living in Cambridge.* [AIP]

## The Season of Particle Accelerators Begins

Rutherford had shown that much physical information could be obtained by bombarding nuclei with $\alpha$ particles emitted by radioactive substances. He is now ready to extend this kind of investigation, by employing different kinds of projectiles to send against target nuclei. Rutherford wonders if is it possible to create a beam of particles and accelerate them, so that they gain enough energy to penetrate the target nucleus. What particles are the best candidates for such an incident beam?

Luckily, the best person to answer his questions is in Cambridge: it's George Gamow. Gamow has no doubt: it is best to accelerate protons, as proton bombardment is certainly more effective, compared to $\alpha$ particles. With two quick calculations *on the back of the envelope,* Gamow explains why, and adds that accelerating the proton beam to an energy of about

1 MeV should provide the particles enough energy to be captured by the target nucleus.

The *Crocodile* is enthusiastic. He immediately involves two of his most expert colleagues, the British John Cockcroft and the Irish Ernest Walton, who are already insiders at the Cavendish Laboratory and can promptly start working.

The experiment will be very successful. In 1932, Cockroft and Walton will build at the Cavendish Laboratory a proton accelerator of 300 KeV (a KeV is one thousandth of a MeV). This machine will turn out to be efficient enough to induce the reaction where a proton hits a nucleus of lithium-7, giving rise to two alpha particles: $p + Li^7 \rightarrow \alpha + \alpha$. In a similar way, they generate a number of other important nuclear reactions.

Einstein considered the above reaction on lithium-7 as the best demonstration of the mass-energy equivalence. Cockroft and Walton will win the Nobel Prize for Physics in 1951, for this class of experiments. Gamow, although he never directly participated to the set-up of the experiments, surely was its main inspirer.

Almost at the same time, on the West Coast of the United States — at Berkeley University — another famous particle accelerator was being built. This one was round shaped — a cyclotron — and its inventor was to become a celebrity too: Ernest Lawrence.

The challenge for the investigation of nuclei and particles by means of accelerators had begun.

## Hoping No New Particle Will Show Up

At the beginning of summer 1930, Gamow's dear friend Lev Landau reaches Cambridge. Switching between research activities in frontier physics and relaxing holidays, *Geo* and *Dau* leave for "a long journey to England and Scotland to visit ancient castles and museums". They travel together on a motorbike that Gamow has just bought from the Birmingham Small Arms Company (BSA), originally a producer of weapons that had turned to motorbikes and cars. Gamow is very proud of his new motorbike: he can take his friends for rides and teach them how to drive. Niels Bohr himself will be one of his most enthusiastic pupils!

At the end of the holiday, Gamow heads back to Copenhagen, where Bohr has got him a scholarship at the Institute for Theoretical Physics for the upcoming academic year 1930–1931. The Soviet Embassy in Copenhagen extends the validity of his passport without any problem. Gamow spends the Christmas holidays with Bohr and other friends in Norway — pleasant and amusing days, as he describes them in his autobiography. Clearly, Gamow and Bohr had become close friends.

Gamow writes his first book: *The Constitution of Atomic Nuclei and Radioactivity* [GG31]. This work represents the very first textbook in theoretical nuclear physics. Gamow believes it is time to systematically collect all available information about this special field of research in one book. However, he is well aware that some of the ideas presented therein are necessarily going to be soon overcome. That's the case of *nuclear electrons*, for instance; but how can one get rid of them, as long as the existence of the neutron is not yet proved?

To highlight some of these too artificial (and soon to be overcome) concepts in the book, Gamow invents a trick: in the text, close to each of them, he prints a special symbol, a sort of alert sign for the reader. In his intentions, these symbols will be useful in the future, whenever new discoveries will require to quickly track and revise ideas in the text. Gamow also wants this special typographic symbol to signal a *high* alert, so he chooses skull and crossbones — indeed, a little too exaggerated, as typical of Gamow. The editor, Clarendon Press, will substitute it with a less gloomy one — a "tilde" (a little wave, ~).

It is worth noticing how Gamow's attitude in writing *The Constitution of Atomic Nuclei and Radioactivity* is similar to Eddington's in writing *The Internal Constitution of the Stars*. Both authors take the calculated risk to write a very complex book that most probably will be outdated within a short period of time.

Gamow's book is ready on May 1, 1931. His friend Landau collaborated with him to the complex mathematical parts.

In 1932, something big happens: the neutron is discovered, with great satisfaction for the entire physical community, and for Gamow in particular. This, together with a number of other important discoveries, now implies that Gamow has to start updating and extending his book, as he had already foreseen.

The new edition is published in 1937, still by Clarendon Press. It is almost twice as long as the first one, and has a new title: *Structure of Atomic Nuclei and Nuclear Transformations* [GG37]. The editor sends Gamow a first copy, accompanied by a letter: "Dear Dr. Gamow: Glad to be able to send you an advance copy of your book. It looks rather stouter and handsomer this time, as benefits maturity. And I hope you are safe from unfortunate radical discoveries in the next year or two" [AIP]. Clearly, the editor is worried that new *unfortunate* discoveries, such as that of the neutron, might once again make this other edition obsolete!

While showing his interviewer Weiner the bunch of old correspondence he had had with Clarendon Press back in time, Gamow is amused and laughs. He also points out that many years later he had written a third version of the book, together with his student Charles Critchfield. The title for the third edition was *Theory of Atomic Nucleus and Nuclear Energy Sources* [GC49], and both Gamow and Critchfield used it as a text book for their classes — Critchfield at the University of Minnesota in Minneapolis, where he had moved to in the meantime. This book was such a success that it was immediately sold out all over the United States.

Upon Gamow's pressing requests to deliver more copies to the US, from Oxford Clarendon Press apologized and expressed their great surprise for the sudden success of a book about nuclear physics. In their letter, they report to Gamow the total amount of copies sent out to the States, and, as an excuse for having underestimated the needed number, they note that "*Unfortunately* the demand for this book has been somewhat heavier than we expected." In a second letter, the Publisher further explains the concept: "We have been very surprised by the demand for books on nuclear physics and an edition which before the war would have lasted us five or more years is gone in as many months, and it is hard to plan in conditions like this."

Gamow must have found this very hilarious.

## Pauli's Magic Intuition: The Neutrino

During winter 1930–1931, Gamow receives from Guglielmo Marconi, the honorary president of the Conference of Nuclear Physics to be held in Rome in October 1931, a formal invitation to give a lecture on the nuclear structure.

The Conference is organized during a phase of great development of particle and nuclear physics, which is in full growth. We here mention two among the most important issues.

A new major issue has appeared at the horizon of physics: the interpretation of the properties of $\beta$ radiation. In this process, an electron is emitted with a continuous distribution of energy, at difference from what happens in the case of $\alpha$ and $\gamma$ radiations: both the $\alpha$ particle and the photon, in fact, are emitted only for discrete values of energy. The first physicist to notice this peculiarity of the $\beta$ radiation is James Chadwick, who had been Rutherford's postgraduate research student in Manchester in 1911. Chadwick makes his discovery in 1914, while he is benefitting from a scholarship at the Physicalisch-Technische Reichsanstalt in Berlin-Charlottenburg, directed by Hans Geiger. Several hypotheses are formulated to explain the energetic spectrum of β emission, including the violation of the principle of energy conservation. A daring idea, that is also supported by Niels Bohr.

In 1930 an important turn takes place: Pauli puts forward the hypothesis that *two* particles are emitted in the β radiation process: an electron — that can be observed during the experiment — together with a second particle, a neutral one with a mass comparable to that of the electron — not observable. These two particles share, in a variable way, the total available energy in the β decay process, and this explains the electron's energetic distribution. On December 4, 1930 Pauli addresses a letter to the physicists gathered at Tübingen for a conference, in order to communicate his idea. He calls the hypothetical particle a *neutron* — a name that Fermi later on will change to *neutrino* (to be precise, in this case it is an *antineutrino*, but this is just a technical aspect).

Pauli talks about his idea at a conference of the American Physical Society in Pasadena, California, on June 16, 1931. The following day the new particle appears on the New York Times. However, Pauli doesn't write any official paper about it yet: he cautiously chooses to wait, to have time to discuss his theory with a number of colleagues, that he is going to meet at the Rome Conference in the coming month of October.

During the 1920s, as a follow up to Rutherford's 1919 results, nuclear disintegration of light elements with α particles emitted by radiative nuclei is investigated not only by Rutherford and his colleague Chadwick at the Cavendish Laboratory, but also by several other experimental groups. Among these, the group led by Walter Bothe at the Physicalisch-Technische Reichsanstalt in Berlin-Charlottenburg, whose results in 1927–1930 seemingly indicate the emission of $\gamma$ nuclear radiation, and thus open up a number of interpretation issues. We will soon discover what that was about.

The 1931 Rome Conference is going to play a fundamental role in the evolution of the most intriguing subjects of physics at that time. It is the first international conference focusing on the newly born subject of nuclear physics. Enrico Fermi is the General Secretary of the Conference: this is an extraordinary occasion for him to get to understand in depth the potential of Pauli's hypothesis on $\beta$ decays, and also to lay the foundations for future investigations to be carried out with his team in Rome, the *ragazzi di via Panisperna.*

Gamow eagerly accepts the invitation to participate to the Rome conference. He also adds a series of pleasant activities to the program of his trip. He plans to spend the summer of 1931 traveling in Europe on his BSA motorbike, join the Rome conference and then travel back to Leningrad via Istanbul and Odessa, to visit his father.

Unfortunately, things don't go according to his plans. In order to travel he needs to extend the validity of his passport, but he is informed by the Russian ambassador that to apply for a visa he must necessarily go back to Russia. Disappointed, Gamow must give up his traveling plans, and also interrupt the flirt with a Danish girl he has just met in Copenhagen.

## 7

# Nothing Particularly Good

*Gamow: So, I remember I went on a trip on the Volga, I guess, and then to Crimea, and then came back to Moscow at the time ... I was told to come two weeks before time I have to depart. And they said, "No, passport is not ready yet."*

*Weiner: Meanwhile you had a specific appointment in Leningrad as a professor?*

*Gamow: We had to get five salaries to get enough money to live. [...]*

*Weiner: How did you split your time, though? Was it possible?*

*Gamow: I was doing nothing. I was trying to get out of Russia.* [AIP]

### No Visa for the Rome Conference

With great regret for having given up both his travel plans around Europe and the Danish girl, Gamow reluctantly returns to Russia. He flies directly from Copenhagen to Moscow, where he must apply for the renewal of his passport, in order to be able to participate in the Rome Conference. Back in Russia, he quickly realizes how radically the situation has changed there, compared to just two years before. Some friends explain that the Soviet regime is now hindering all kinds of exchanges between Russian and foreigner scientists. According to Stalin's vision, in fact, there must be a clear distinction between *capitalistic* and *proletarian* science, in a quite scary analogy with what was happening in Germany, with Hitler's separation between *Jewish* and *Aryan* science. Given the circumstances, Gamow's friends are very surprised that he decided to come back to

Russia; clearly, being abroad, he was not at all aware of the ongoing drastic changes.

Gamow's problems begin when he contacts the Narkompros to apply for an official authorization to go to Rome. The release of his passport is continuously delayed, a day after the other. As the date of conference is approaching, Gamow starts fearing he will not make it in time. He writes directly to Marconi, asking him to postpone as much as possible the day of his lecture and, in the meantime, he keeps pressing the Russian bureaucrats to speed up the procedure. But all his efforts are in vain. In the end, he is even advised to withdraw his passport request!

The possibility to participate in the Rome Conference thus vanishes. The scientific report about the quantum theory of nuclear structure that Gamow had prepared for his lecture at the event, is read on his behalf by his friend and colleague Max Delbrück. Many of the physicists attending the conference send Gamow a postcard from Rome, to express their solidarity.

During his interview with Weiner, Gamow shows this postcard, carefully indicating, one after the other, some of the most important signatures on it: Marie Curie, Compton, Millikan, Blackett, Heisenberg and many more.

Among these signatures, one is in Cyrillic: it's Gleb Vasilievich Wataghin's, and Gamow immediately recognizes it. Wataghin is a physicist, born a couple of years before Gamow in Birzula, Ukraine, just about 200 km from Odessa. Wataghin and his family emigrate to Italy when Gleb is 18 years old; between 1922 and 1924 he graduates in Physics first and then in Mathematics at the University of Turin. He gives courses at the Application School of Artillery and Engineering, at the Polytechnic School and at the University of Turin. He obtains the Italian citizenship in 1929. Extremely talented, Wataghin doesn't go unnoticed within the scientific community, and he is chosen to collaborate with Enrico Fermi at the Rome Conference as a scientific secretary. Before holding the chair of Experimental Physics in Turin in 1949, between 1934 and 1948 Wataghin is in Sao Paulo, Brazil, where he is engaged in a big project regarding the creation of a new faculty. It is precisely during these years that Wataghin organizes some cycles of conferences for Gamow in Brazil. We will go back to Gamow's 1939 visit to Brazil and to its important scientific implications further on in our book.

Gamow and Wataghin become very close friends, and their friendship will last their entire lives: besides being both Ukrainian emigrants, they share the same scientific interests, as well as remarkable social skills.

The schedule for the Rome meeting clearly bears the stamp of Enrico Fermi, who, as general secretary of the conference, is in charge of the scientific program. His choice is to limit the number of official reports and give space to young emerging physicists; he also wants to leave as much free time as possible for informal discussions.

Fermi's formula will turn out to be very successful. This is how Samuel Goudsmit, a renowned scientist from the University of Michigan, will refer to the Rome Conference in a letter to science historian Laurie Brown: "It was the best organized meeting I ever attended, because there was very much time available for informal discussions and get-togethers [...]. Fermi ordered the then 'young' participants, namely [Nevill] Mott, [Bruno] Rossi, [George] Gamow (who could not leave Russia, but sent a manuscript) and myself, to prepare summary papers for discussion" [Br].

Goudsmit himself gives one of the most interesting lectures in Rome, in which he reports the content of the lecture Pauli had given earlier in June that same year, at the meeting of the American Society, in Pasadena, California. Pauli, even though attending the Rome Conference, never speaks openly about the neutrino, but shares his thoughts only in private conversations.

Besides offering an excellent scientific program, the Rome Conference is enriched with a wide schedule of social events, sponsored by the *Società Italiana Edison di Elettricità* (Italian Electric Company Edison) through the Volta Foundation, in connection with the *Accademia d'Italia* (Italian Academy). The reasons behind the huge economical investments are clearly political. The fascist regime was in search of visibility at an international level, in order to somehow cover the shame of the dictatorial system. Mussolini himself was a great promoter of the organization of the conference in Rome and gave the inaugural speech. Just a few years later, the fascist regime in Italy evolved in a direction that led to the promulgation of the racial laws, as well as to a number of other well- known tragedies.

In 1956, in a letter to Franco Rasetti, Pauli shares his thoughts about two memorable experiences he had had in Rome: "*Horribile dictu*, I had to shake hands with Mussolini" and "Fermi asked me to talk about my new idea [the neutrino], but I was still cautious and did not speak in public." These memories, as well as other countless curiosities, are reported by Abraham Pais in his extraordinary book *Inward Bound* [Pa].

Enrico Fermi had to act very smartly in order to organize a conference of great scientific importance and at the same time try to limit as much as possible formalities and celebrations, of which Guglielmo Marconi (who was pretty well integrated in the fascist regime) was in charge of.

## In the Meantime, New Particles

Having missed the Rome Conference causes Gamow great disappointment, especially because he feels isolated from the scientific community during times of extremely rapid developments for 20th century physics. This is what was going on in the meantime.

Immediately after the Rome Conference, Marie Curie, impressed by the interesting results obtained by Walter Bothe and colleagues in Berlin, suggests her daughter Irène and Frédéric Joliot, in Paris, to start investigating the effects of bombarding light nuclei (beryllium and boron) with $\alpha$ particles, emitted by a source of polonium. The analysis of their own results, lead the Joliot-Curie in January 1932 to conclude that these nuclei, when bombarded with $\alpha$ particles, emit $\gamma$ radiation, which in turn can penetrate materials placed around the target, and generate protons. Their interpretation, however, is based on a number of hazardous theoretical hypotheses.

The Joliot-Curie's conclusions are very weak, and soon proved wrong by James Chadwick. At the Cavendish Laboratory, Chadwick repeats the same experiments the Joliot-Curie had studied in Paris, and shows that what is being emitted by the bombarded nuclei is not $\gamma$ radiation, but rather a new particle, with a mass equivalent to that of the proton, but with no electric charge: the *neutron*. This particle had already been predicted by Rutherford in 1920, with great foresight. Chadwick first announces his extraordinary discovery in *Nature* in February 1932 [Ch].

This result put a definitive end to the Joliot-Curie's incorrect interpretations: they simply never realized they had produced a neutron. Ettore Majorana seemed to have well understood the misinterpretation when he, as reported by Emilio Segrè, sarcastically commented the Joliot-Curie's results as follows: "*Ah, guarda questi idioti — hanno scoperto il protone neutro, e non l'hanno neppure riconosciuto.*" ("look at these idiots — they have discovered the neutral proton and didn't even recognize it.") [Se].

The discovery of the neutron is of fundamental importance in the development of nuclear physics. Finally, the atomic nucleus can be described for what it is — a system composed of protons and neutrons, without unnecessary complications such as that of *nuclear electrons.*

In September 1932 one more new particle shows up, this time thanks to an experimental research carried out in the United States. At the Californian Institute of Technology, Robert Millikan wishes to confirm his theory on cosmic rays, and asks his student Carl Anderson to set up an experiment, aimed at visualizing the traces of the passage of charged particles (with a technique that had already been applied by Dimitr Skobeltzyn in Russia). Anderson finds evidence of the passage of particles that are similar to electrons, but with an opposite electric charge — that is, he had demonstrated the existence of the electron's antiparticle, the *positron*, exactly as predicted theoretically by P.A.M. Dirac's quantum theory.

In the meantime, Pauli's idea on the existence of the neutrino seems to become more and more consistent. In July 1932, during a conference in Paris, in fact, Fermi openly states that some features of $\beta$ decay can be quite simply explained in terms of production of a neutrino. Fermi probably has already in mind, at least roughly, what he will develop in detail by the end of the following year: the quantum theory of $\beta$ decay. He is still further investigating the mathematical tools he needs in order to give a complete formal structure to his physical intuition.

Gamow does not partake to any of these important events, distracted as he is by the bureaucratic issues he must solve. He feels also isolated from the international scene. When Weiner asks him if he remembers anything in particular about the discovery of the positron and what had been his reaction, Gamow replies with words that clearly describe how he must have felt during that period: "I was in Leningrad. [...] I just read it in *Nature*, apparently. [...] I don't remember. Probably I was too busy worrying about which frontier of Soviet Russia is easiest to cross. I didn't do much physics in Russia at this time."

In 1931–1932, forced to remain in the Soviet Union, Gamow holds a position as professor at the University of Leningrad. About this period, in his autobiography Gamow expresses very bitter considerations regarding the oppressive regime imposed by Stalin to the world of culture and its devastating consequences. In the field of biology and genetics, for instance, the regime will support the ideas of Trofim Lysenko, an agronomist who will become a *champion of the Stalinian cultural revolution.*

> Gamow had personally experienced the regime's methods just a few years earlier, in Leningrad. He was involved in a quite emblematic episode, that touched him deeply, as described in his autobiography. One day, he and Landau were discussing about physics at the famous Borgman Library, when their friend Bronstein suddenly interrupted them: looking very amused, he showed them an article that had just been published in the Soviet Encyclopedia. It was signed by comrade Gessen, whom they all knew very well, as he was the political supervisor at the Physics Institute of Moscow, that is, the person in charge of assuring that every kind of research carried out at the Institute complied with the cultural directives of the Soviet regime. As an example, the *weird ideas* preached by Einstein or Heisenberg were considered unacceptable, because the regime

did not tolerate that the Indetermination Principle set a limit to the possibility to localize a particle nor that a particle could never overcome the speed of light! In the article Bronstein was showing them, comrade Gessen proposed once again the existence of the *luminiferous aether*, in contrast with Einstein's too presumptuous ideas.

One can easily imagine how Bronstein's hilarity was immediately shared by both Gamow and Landau. The three of them, together with two PhD students, decided to make up a proper mockery, to make fun of the article. They sent Gessen a telegram, in which they stated that his article had stimulated them to start an investigation to demonstrate the existence of aether. Not only, they pushed themselves so far as to ask Gessen to be their guide in the study of this extraordinary caloric and electric fluid. The joke had a disruptive effect! Immediate consequences of their action followed: people were fired and scholarships suspended. Gamow was the only one to remain untouched by the regime's revenge, since his position was not under Gessen's jurisdiction.

Between 1931 and 1933 Gamow and friends become very active again. Gamow has now gained a certain political weight, and has been elected to Corresponding Member of the very prestigious Soviet Academy of Sciences. He is the leader of a group of people called the *Young Turks*, who claim official recognition by the Soviet institutions of their research plans. Gamow, Landau and Bronstein are together in this fight, even though their political orientations are completely opposite: Gamow is very critical towards the regime, whereas both Bronstein and Landau are initially supporters — but will turn out as victims of the Stalinist Great Purges, in the dramatic follow up.

In 1937, in fact, Bronstein will be arrested and sentenced to death, in a typical mock trial: he is executed on the same day of the sentence, in 1938. Landau will be arrested in 1938, after being coerced to confess having conspired against the state. He will be set free a year later, in 1939, thanks to the intervention of his friend and colleague Peter Kapitza.

Even though tragedies of this kind could not even be imagined back in 1932, Gamow feels oppressed by the political situation and comes to the conclusion that the only way out is fleeing his Country.

And this is when Gamow's autobiography turns to storytelling: he starts a quite accurate and very amusing description of all his various, and at times extravagant, escaping plans — and of their unsuccessful outcomes [GG70].

## Fleeing is the Only Way

Something else has happened to Gamow in the meantime: he is not making plans to leave Russia alone anymore. During the days he spent in Moscow trying to arrange his passport papers, he has met a girl, Lyobov Vohminzeva, that he calls Rho, a Physics graduate from the University of Moscow. They fall in love and will marry within very short. From that moment on, they figure out together by which routes and how they can eventually try to flee Russia illegally.

According to Gamow's story, Geo and Rho's first idea is to try to cross the Russian western border. However, evaluating this possibility in detail, they realize that the main challenge is represented by the very arduous mountain passages they will need to cross and by the thorough controls at every crossing point. So, Rho suggests, why not trying to escape by boat instead, crossing the Black Sea? This will actually be their first real attempt, that Gamow ironically refers to as his *Crimean Campaign*. The idea is to cross the Black Sea during the summer of 1932: sailing from a small location near Yalta, on the Crimean Peninsula, and heading south, in direction of the nearest spot on the Turkish coast — that is, a southward route of about 270 kilometers. The boat they choose is a kind of unstable kayak, endowed with paddles. Hoping the weather will hold, crossing time is estimated in 5–6 days.

How much of this story is true and how much Gamow made up, is hard to tell. One may wonder if he just wanted to show off his writing skills, or if it is actually meant to be a credible story. Anyhow, his description is so rich in details that it occupies several pages in his autobiography. Ready to leave Russia forever, George and Rho sail on a sunny summer day and on calm water; but the pleasant trip soon transforms into a struggle to survive, when a sudden storm forces them to find shelter on a beach near Sebastopolis — they end up just a little west of their departure spot!

Not at all discouraged by their first failure, Gamow starts making new fleeing plans: this time, he wants to cross the northernmost border of Russia, beyond the arctic polar circle. It's a region that he knows quite well, where the borders of Russia, Finland and Norway entangle. He is now considering crossing either by sea or by land, starting from Murmansk, on the Kola Peninsula, or from Khibiny, in the Russian Karelia.

In order to finalize the details of their plan directly on spot, in the summer of 1933 he and Rho spend a week at the KSU base (the Commission for helping scientists). Unexpectedly, they are accompanied on this vacation by their friend Landau, a convinced Marxist with no desire to escape. At the end of the holiday, Landau takes a train back to Leningrad, whereas Gamow continues his trip to Murmansk together with his wife. Their attempt will fail again, as it turns out that all the possible escaping routes are impracticable.

What remains of this journey to Khibiny is a quite important scientific paper, that Gamow and Landau write together: in a note they send to *Nature* [GL], they discuss an application of Atkinson's and Houterman's work — the before mentioned one on thermonuclear reactions in the stars. This note, entitled *Internal Temperature of Stars,* is a first evident sign that Gamow is now progressively leaving behind pure nuclear physics and moving towards its applications to astrophysics.

Once Gamow fully realizes that leaving Russia illegally is an impossible task, something almost miraculous happens. The Narkompros informs Gamow that he is invited to participate to the 7th Solvay Conference, that will be held in Brussels in the coming month of October, as the official Russian delegate. And that in order for him to participate, he will be issued a visa.

Gamow's invitation to Brussels is not fortuitous: Niels Bohr is behind it. Having realized that Gamow has been isolated from the rest of the scientific community for too long, he wants to rapidly reintegrate him, also considering the crucial phase that physics is undergoing. The Solvay Conference is absolutely the best opportunity to drag him once again on the scene.

The title of the Conference is *Structure et propriétés des noyaux atomiques* (Structure and properties of atomic nuclei): the topics are obviously of direct interest for Gamow. However, in his autobiography he reports very briefly the scientific aspects of his visit to Brussels: his mind is clearly elsewhere, mainly concentrated on figuring out how to avoid returning to the Soviet Union at the end of the Conference. It was certainly not an easy decision to make: not going back wouldn't simply mean leaving his Country as a normal citizen, it had to do with defecting Russia after having been its official representative. A circumstance that, under the Soviet regulations of that time, would have implied death penalty.

Despite his "difficult psychological situation", Gamow decides not to return to his motherland at the end of the Conference. Next step is then making sure that his wife is allowed to join him on his journey to Brussels.

Obtaining a visa also for Rho is definitely not easy. But Gamow, as usual, doesn't give up and decides to go personally to Moscow to talk directly to the President of the Soviet Council of People's Commissars, Vjačeslav Molotov. During his interview with Molotov, Gamow talks very frankly and confidentially. "Since my wife is a physicist, I could claim that I need her presence at the conference as a scientific secretary — he more or less says — but this would not be the truth. The point is that my wife has never been abroad, and I want to take her to Paris afterwards, show her Paris and buy her some clothes". All this makes grim Molotov smile, implying that a solution might be found.

Actually, things will not be as simple as they sounded. The Russian bureaucracy is once again very slow and Gamow has to threaten that he will not represent Russia at the conference. In the end, he receives both visas, for himself and for Rho.

Busy as he is trying to speed up the regime's bureaucracy on one hand and preparing escaping plans on the other, during the years between 1931 and 1933 Gamow has almost no time left to dedicate to scientific research. Luckily, his economy is safe thanks to five different salaries he receives, corresponding to five different didactic and scientific assignments. Besides being professor at the University of Leningrad, in fact, he is a member of the Physical Research Institute, of the Physico-Mathematical Institute and of the Radium Institute, and he also works as a consultant at the Ioffe Institute.

To the question Weiner asks him as to how he managed to split his time during his *Crimean Campaign*, Gamow's answer is quite sharp and revealing: "I was doing nothing, I was trying to get out of Russia". This remark, even though it sounds somewhat exaggerated — in Gamow's style — was pretty much true. Gamow's scientific production in 1931–1933 is quantitatively poor, although very rich in terms of quality. As an example, in the report he had prepared for the Rome Conference, he touches upon topics in nuclear physics that many years later will lead to the formulation of the nuclear *shell* model, thanks to the work carried out by Maria Goeppert-Mayer and Hans Jensen.

With the newly issued visas finally in his hands, Gamow quickly organizes the trip to reach the Solvay Conference, together with Rho. They leave from the *Finland* train station in Leningrad, heading to Helsinki. From there they continue their journey via Sweden and Denmark and finally reach Brussels, on a route that avoids Germany. Gamow must have been in a very good mood. On the afternoon of the last day he will ever spend in his motherland Russia, he takes Rho to the theater, to enjoy the Koniok Gorbunok ballet.

# Destination: United States of America

*Weiner: What about your paper there [at the Solvay Conference]?*

*Gamow: Pauli told me that I did a horrible paper there — he didn't like it at all. [...] It really wasn't much. It was kind of fitting experimental data.... Well, that's what I was doing in Russia; I wasn't doing much.*

*Weiner: In that letter you wrote to him [Goudsmit] you talk of the land of the skyscrapers and the atom-splitters. Now a lot of people would be surprised at that reference, of atom-splitters.*

*Gamow: Well, this is, of course, Lawrence.*

*Weiner: It's a reference to Lawrence's work. Was Lawrence's work pretty well regarded or was it looked down on in Europe?*

*Gamow: Well, you see, Lawrence is a definite case. [...] He just didn't give a damn what the nuclei do; he just wanted to smash them. So he did. And there are such people. [AIP]*

## At the Solvay Conference, with His Mind Elsewhere

In 1933, the 7th Solvay Conference in Brussels represented one of the most important events for the scientific community. Although the Rome conference just two years earlier already focused on nuclear physics, this new frontier topic is so important, that also the Solvay Conference is entirely dedicated to it. The organizers apply the same strategy Fermi had chosen for the Rome Conference, that is, they involve as many young scientists as possible.

Paul Langevin is the president of the scientific committee. He is firmly convinced that "A young physics requires young physicists".[1] Besides the younger scientists, however, all the most renowned physicists of that time gather at the Conference, with one exception: Albert Einstein, who has just left Europe and moved to the United States. The schedule is very rich and practically every aspect of nuclear physics is touched upon.

One of the main subjects of discussion is particle accelerators physics: John Cockcroft shows the results he has obtained with the proton accelerator he has built together with Walton at the Cavendish Laboratory, and Ernest Lawrence shares the experimental research he is carrying out on Berkeley's cyclotron. The atmosphere is very lively and when it comes to discussing the value for the mass of the neutron, it even heats up.

The other big topic of discussion is, of course, β decay. At this conference Wolfgang Pauli finally talks about his theory in public: he advocates the existence of the neutrino, and defines in detail the properties that this new particle, that he had suggested three years earlier, should have. Pauli once again declares that a violation of the energy conservation principle, as suggested by Bohr, to justify the electron energy distribution in β decays, is definitely unacceptable.

Although present at the conference, Fermi does not partake in the discussion about β decay; he will give the final touches to his theory immediately after, and send his well-known note to Nature in December 1933. In the following couple of years,

---

[1] P. Langevin, Institut International de Physique Solvay (1934), p. x, as reported by R. Stuewer [St].

Fermi will fully engage himself in his famous experiments with neutrons, thanks to which he will win a Nobel Prize for Physics in 1938.

The scientific report that Gamow gives at the Solvay Conference does not seem to have been a particularly interesting one, and years later he will talk about it quite coldly. He had probably prepared his speech in a hurry and just to be able to send it on time to Brussels directly from Russia, as required by the organizers.

The topic Gamow presented was the electromagnetic radiation emitted by an atomic nucleus during transitions from a quantum state to another. "My paper was actually experimental; it was zoology, as Einstein would call it, of gamma rays, fitting them into the levels — it was nothing especially exciting" and adds "Pauli told me that I did a horrible paper there — he didn't like it at all" [AIP].

Gamow felt actually very little involved in the Conference itself. When Weiner asks him "Do you remember any special point of interest in the discussion, in general, at the meeting — I don't mean necessarily in your paper. Do you remember what was the highlight of the meeting?" Gamow's first answer is "Nothing special". And then, reconsidering the thing, he says "Well, the highlight was essentially Cockcroft and Walton because they just broke the lithium by artificial protons. [...] Oh, Lawrence and new results ... In any case, nuclear reactions, cross-sections and things like that. There was, as I remember, very little theory" [AIP].

Gamow's quite cold reaction is not surprising, after all. At the time of the Solvay Conference his mind was elsewhere, mainly concentrated on finding a way to put into practice his plan to flee Russia forever. But in order to do so, he needed to act very cautiously.

Based on the confidential relationship he had to Bohr, at the end of the Conference he shared with him his intention not to return to Russia. Bohr's first reaction was "But you can't do that!"

From here onwards, the story continues according to Gamow's own description, as he puts it in his autobiography. Bohr admitted having personally suggested Gamow's name as the Russian delegate to the Conference, in order to give him a chance to get involved again in the scientific community. He also tells him that it had been made possible only thanks to Paul Langevin's personal contacts, since he chaired the Franco-Russian Scientific Cooperation Committee. This explains Bohr's first reaction to Gamow's confidence about his plans: obviously, Langevin had to be informed about Gamow's intentions to defect the Soviet Union.

The occasion for Gamow to finally solve this political and diplomatic problem comes at the end of the Conference, when he and Rho go to Paris. During a dinner at Marie Curie's house, Gamow confesses to her that he is feeling morally obliged to go back to Russia, against his own will. Marie Curie heartens him, and promises she will personally talk to Paul Langevin about this delicate matter. Gamow must have realized in that very moment that his problem was about to be solved. Marie Curie's intercession had very good chances to be successful, especially thanks to the very close (and gossip-topic) relationship she had with Paul Langevin. And in fact, it was. After her chat with Langevin, Marie Curie meets Gamow in the corridors of the Sorbonne, she puts her hand on his shoulder and laconically says: "*Gamow, vous resterez ici*" (Gamow, you will stay here).

At that point, young Gamow (he is not thirty years old yet) has not only managed to be officially accepted by the French community of physicists, but to be considered as one of them.

The umbilical cord between Gamow and his motherland is finally cut. Abram Ioffe, the director of the Physical-Technical Institute in Leningrad, who had been attending the conference in Brussels, is informed about Gamow's decision and makes sure his salary is immediately suspended, retroactively from the day he had left Russia.

Clearly, Gamow has now become a *persona non grata* to the Soviet Regime. This is quite obvious also from a number of minor details: as an example, in a photograph taken back in 1929 in Leningrad, portraying a group of young physicists around Professor Frenkel, Gamow's figure is, years later, substituted manually with a fake pillar.

## USA: East Coast or West Coast?

Gamow must now make another important decision: what will his next destination be? Once again, it is Niels Bohr who provides the best advice. No doubt, he suggests him to try to get to the United States. Winds of war are blowing in Europe, and the USA are certainly in need of expert theoretical physicists.

Following his advice, Gamow writes a letter to Sam Goudsmit, at the University of Michigan. A year earlier, in 1932, Goudsmit had invited him to give some classes at Ann Arbor's summer school. "I wonder whether the kind invitation to lecture in Ann Arbor, which I got last summer and could not follow in spite of all attempts, could not be postponed on this summer", Gamow asks, and adds that "Now nothing except the collision with an iceberg in Atlantic Ocean could prevent my arrival. [...] I am looking for to see the skyscrapers and atom splitters of USA."

In his long letter, Gamow expresses his desire to move to the United States for a longer period of time, mentioning he had previously discussed it with Lawrence. Gamow is actually quite sure he will receive an offer from the University of California in Berkeley. So much that, in that same letter, he informs Goudsmit about his plans to cross the United States from East to West, heading to California, at the end of Ann Arbor's summer school.

Probably, Gamow had not talked directly to Lawrence about his future: Niels Bohr had, on his behalf. After all, one of Bohr's main skills was sponsoring young talented physicists for various positions in institutes all around the world, sometimes even using small tricks. That was the case, for instance, of Viki Weisskopf, a young physicist who will become a very renowned scientist. Bohr had written to the University of Rochester, asking them to offer Weisskopf a position. Although very politely, they gave a negative reply. Still very determined to reach his goal, Bohr then wrote back — thanking them very warmly for their offer. In Rochester they couldn't do anything but accept their new guest. Certainly, they soon realized it had been a very good choice!

Things didn't turn out as simple for Gamow. Bohr had actually talked to Lawrence and felt quite optimistic, especially because he knew that theoretical physicists were lacking at Berkeley's cyclotron. Edward Teller had spoken to Lawrence too, to make sure that Gamow was offered a position there.

But Teller, Gamow says, "told me he had run into a complete wall because I think they said that ... I can think, I can think, I can talk, but I cannot work. And Lawrence wanted to have a good worker. It was something like this." [AIP]

Considering that Gamow refers to particle accelerators as *atom splitters*, and ironically blames experimentalists for being too focused on the splitting itself, rather than on understanding nuclear properties, he probably wouldn't have felt at ease in Berkeley anyway.

One more issue made things further complicated. Berkeley, as well as other universities across the Country, had faced huge budget problems during the 1930s, as a consequence of the economic cutbacks during the years of the Great Depression. New positions were therefore extremely hard to get.

Gamow will receive a joint offer from the George Washington University and the Carnegie Institution, that perfectly fits his interests and desires. This offer opens up for Gamow the possibility to fully develop his scientific career, independently.

Finally, things are set and Gamow's short-term program now consists in going to Ann Arbor for the 1934 summer session. What still must be covered is the upcoming winter term, before moving to Washington in 1935. His European colleagues take care of this, splitting Gamow's free months on three different locations: Cambridge, Paris and Copenhagen. Such proofs of solidarity from his friends are clear signs of how much Gamow is appreciated. The economic return of these short assignments is quite small, and certainly not enough to cover the costs for the ship tickets to America he must buy for himself and for Rho. Once again, he will receive precious help from his friends — Bohr and Rutherford will lend him the money he needs.

> Gamow clarifies in his autobiography that as soon as he settled in the USA and had earned enough money, he paid back both of them. Which might sound pretty normal and uninteresting, if it wasn't for the circumstances. He had gone to the post office and filled up a money order each, to be sent to the two famous physicists. The postal worker registered the orders obviously not recognizing the names of the recipients. A real pity, Gamow says, not having kept the two original receipts.

In terms of physics, that period was not a particulary fruitful one for Gamow. Despite the personal problems he is trying to solve, during these years he manages anyhow to lay the foundations of a research that he will complete together with Teller, once he is fully settled in the United States. It has to do with a particular kind of β transitions, the ones that go under the name of *Gamow–Teller transitions.* A preliminary paper about this, is published in the Proceedings of the Royal Society: *Nuclear Spin of Radioactive Elements* [GG34]. One more milestone on Gamow's scientific path is placed.

9

# I Like to Be a Pioneer

*Weiner: You mentioned that you got bored with nuclear physics because it got complicated [...].*

*Gamow: I started nuclear physics because in 1928 everybody was doing atomic and molecular structure [...]. I didn't want to get mixed up with all this, so I decided to choose myself a corner where nobody was doing anything, so I chose nuclear physics. And in time nuclear physics blew up into a big thing, so I moved to nuclear astronomy, to nuclear astrophysics, cosmology [...]. I like the pioneering thing. I would rather go in these mountains [Colorado] than in California. I like these mountains much better than California, where they have a hot dog stand on the top of each mountain. [AIP]*

## On What Conditions Accepting a Position in Washington

Not having collided with any iceberg in the Atlantic Ocean, in the summer of 1934 Gamow reaches Ann Arbor. Time has come to start getting familiar with the American environment.

The invitation he received to give a course is quite emblematic of how much Gamow was already renowned and respected in the new continent. The theoretical physics summer schools in Ann Arbor had been conceived to improve the preparation of American physicists in theoretical aspects. Therefore, the best available teachers were required, in order to attract as many students as possible.

During that summer of 1934, the pool of teachers invited to Ann Arbor includes, besides Gamow, Werner Heisenberg and Robert Oppenheimer. During his interview, Gamow mentions Oppenheimer among the many people he had met in Ann Arbor. "I was enchanted", Gamow says about him, "he was interested in Persia, he knew Sanskrit, and all sorts of things, poetry ... Yes, I think there were only two: Robert and me". Weiner wonders whether Gamow had discussed some physics with Oppenheimer during the time they spent together, but Gamow's answer is simply "Oh, I don't think so" — which sounds quite unbelievable.

Gamow also reminds having been assigned two students, to assist him during his stay in Ann Arbor, sort of *batmen*, as the officers' personal attendants are called in the British army. And he still remembers their names: Kemble and Tompkins. They were his personal chauffeurs and guides, and at times even sparring partners in some sports — Gamow liked swimming and playing tennis. Gamow doesn't know why, but Tompkins' name sounds amusing to him, so much that he will choose it to designate the protagonist of a series of popular books he will write many years later. We will meet Mr. Tompkins further on in our book.

Under the insistence of the interviewer, Gamow finally also mentions the names of other renowned physicists he had met in Ann Arbor during the summer of 1934: the organizers of the summer school, Samuel Goudsmit and George Uhlenbeck, both Dutch-American citizens, and Arthur Compton.

About Compton, who had won a Nobel Prize for physics in 1927 for the discovery of a physical process that bears his name, Gamow recalls that he was a good tennis player and that once, during a match, he had "intentionally" shot a ball hitting him in the genitals. A quite funny detail to remember.

It is precisely during that summer school in Ann Arbor, that Gamow receives the letter with which Cloyd Heck Marvin, the president of the George Washington University (GWU) in Washington D.C., offers him a professorship in physics.

No doubt, it was the proposal he was waiting for. After his unsuccessful attempts to go to Berkeley, his plans to stay in the United States could have been realized only if he had received a proposal of this kind. It was probably not even important where it came from. What was really crucial for him and for Rho was to be given an opportunity to settle for some time in America.

The reasons why the GWU decided to offer Gamow a position, were mainly two. On one hand, professor Marvin wished to empower the departments of physics, chemistry and biology at his institute, and in order to do so, he needed to introduce new courses and new teachers. On the other hand, the GWU collaborated with the Carnegie Institution in Washington, directed by Merle Tuve, who at that time also directed the Department of Terrestrial Magnetism. Tuve was looking for a theoretical physicist, expert in nuclear physics, in order to support the studies of the proton-proton collision processes he was carrying out on the particle accelerator available at the Institute. However, he could not afford hiring a full-time theoretical physicist. So, because of the existing collaboration with the GWU (of which Tuve was a counselor), the two institutes decided to make a joint proposal to Gamow: a position as professor at GWU and as a (free) consultant at the Carnegie Institution. Gamow accepts, enthusiastically. This offer allows him to stay in the USA for at least five years and projects him straight into the world of research in America, in a very privileged position. After all, he had been chosen because he was an innovator and a pioneer in various fields of nuclear physics.

In accepting the proposal, Gamow sets three conditions:

- that he is given the possibility to invite to the GWU another physicist of his choice, in order to have a constant reference person to confront his ideas with. He even suggests the name: Edward Teller, who at that time was in England;
- that he is supported in the organization of annual conferences on theoretical physics, to be held at the GWU, modelled on the ones Niels Bohr used to organize in Copenhagen;

- that his role is that of "visiting professor", in order to give his new position a temporary profile and thus avoid raising suspicion among soviet authorities about his definitive defection.

In this particular phase of his life, Gamow's personality is definitely changing. So far, we have encountered a young and brilliant George Gamow, a very unusual scientist. Intuitive and ingenious, this young man is always ready to throw himself into new adventures, and fearlessly challenge the world. Now, facing a new phase of his life, a new side of his character is emerging: he is a more mature and aware man, who makes plans for his future in a more organized way and with specific goals. He now feels he has gained both the authority and the freedom to play a determinant role in the development of theoretical physics in the New Continent. The three conditions Gamow sets before accepting the position at the GWU, reveal some important aspects. His request as to invite Edward Teller to join him, implies his need to have a trustworthy colleague with whom to confront. The two had met in Europe years earlier and immediately got along. Gamow needs to discuss every new idea with a colleague before elaborating it in detail, and Edward Teller would have been the perfect sparring partner in this. Besides, Teller could have become a precious partner in the organization of the planned annual conferences.

These conferences represented a very important part of Gamow's program. Perhaps, always a step ahead everyone else, he had already realized that the new frontiers of research consisted in sharing and confronting results, and that the free exchange of ideas was an essential premise for the birth of new and better ones. Shaped on the models of the meetings held in Copenhagen and in Rome, and on the Solvay Conference, such conferences could be replicated in America, in a more modern key. The promotion of annual gatherings of the biggest names in physics had become something crucial for Gamow, also because these meetings represented the possibility to dismantle an old-fashioned mentality, according to which scientific results were somewhat nationalistic records (just think of what was happening in Germany in the 1930s). Gamow wanted to introduce a more modern approach to scientific research, as a common front of cooperation. In this, Gamow was definitely a forerunner of something that will need many more years to be achieved.

Finally, the cautiousness with which he requires his position to be defined as *temporary*, describes a man whose goal is clear: he is determined not to go back to Russia, but at the same time is well aware of how dangerous it is to stand against the Soviet regime during such a delicate historical phase.

The George Washington University accepts all of Gamow's conditions. This will represent a fundamental turning point for the evolution of physics, in America and elsewhere.

Edward Teller's arrival at the GWU did not only enrich the staff of the department with an important scientist — something that certainly was in president Marvin's plans. His presence in the United States will within short become essential for the Manhattan Project and the development of atomic bombs.

The Washington Conferences of Theoretical Physics were organized annually from 1935 to 1942 — preferably in the month of April ("because that's when cherry trees bloom", Gamow explains). Gamow, Teller and Tuve were responsible for the topics of each conference, as well as for the list of participants. Their idea was to gather around 30–40 theoretical physicists, to discuss a specific field of research. Neither a paper before, nor a report after were required: the conference was intended to be an occasion for sharing results and discussing the developments within the chosen field.

Tuve underlines that, as these were "conferences for theoretical physicists", experimentalists (as himself) were allowed to participate only as *audience*, as they had no theoretical contribution to share. It's funny how a sort of competition between theoretical and experimental physicists is historically rooted!

Among the names of theoretical physicists that gathered at the GWU during the years are Enrico Fermi, Linus Pauling, Niels Bohr, Eugene Wigner, Hans Bethe, John Wheeler, Subrahmanyan Chandrasekhar, J. Robert Oppenheimer, Leo Szilard, Richard Feynman, Julian Schwinger, Hermann Weyl, Martin Schwarzschild. Albert Einstein does not appear in this list: by his nature, he did not like to attend conferences, and the early century Solvay meetings had represented his only exceptions.

### An Inseparable Couple: George and Edward

The subjects of the annual conferences in Washington were chosen according to the fields of interest of their main organizers, Gamow and Teller. Therefore, the first conference organized by GWU, in 1935, was obviously dedicated to nuclear physics. Gamow and Teller were at that time conceiving the first ideas that will lead them to formulate a generalization of Fermi's original theory of weak interactions and to develop the study of the before mentioned *Gamow–Teller transitions*.

We had already reminded that Fermi, in one of his fundamental articles in 1933 [Fe33][1] had laid the foundations of the theoretical explanation of (negative) $\beta$ decays, as the result of the transformation of a neutron into a proton inside a

---

[1] It is well known that Fermi had sent his article first to *Nature*, but it had been rejected.

nucleus, with the emission of a couple of light particles: an electron and an (anti)neutrino. Fermi had intuitively understood the correspondence between the emission of an electron-antineutrino couple in nuclear $\beta$ decays and the emission of a photon from a nucleus, when it undergoes a transition from a higher to a lower energy state. With this observation, Fermi practically established the prerequisites for a unification of weak and electromagnetic interactions. This idea will then be further developed into the Standard Model of particle physics, at the end of the 1960s.

Fermi's pioneering work on weak interactions did not take into account the effects due to intrinsic angular momentum, that is, of *spin* of the particles involved. This is precisely what Gamow and Teller investigate, demonstrating the importance of the role of spin, especially in the case of a decaying nucleus that undergoes a transition with a *change* of intrinsic angular momentum — in such cases the *Fermi transition* is absent.

The collaboration between Gamow and Teller starts working perfectly and their *modus operandi* is quite unique. Something Teller will remember in his older days, are Gamow's early morning phone calls — Teller's *early* morning, that is after 9:30 — to inform him about some new idea. "And that idea was simply wrong. Almost always! Geo Gamow had the wonderful property that he did not mind being wrong. He did not do it for the prestige. He did it for fun. He did it for love. And when his idea was not wrong, it was not only right, it was right and new" [HPA].

According to Gamow, the two had their own daily routine. Gamow drove his son to school, then showed up at Teller's house to spend the morning in long conversations about physics. After all, Gamow was completely free to organize his time as he preferred, because the classes he was teaching at the University were concentrated in just two days a week, on Tuesday and Thursday evenings. This schedule fit both Gamow's and his students' needs, as most of them worked during the days.

So, a discussion after the other, Gamow and Teller slowly conceive the idea that nuclear physics can be applied to astrophysics. Gamow actually had already shown his great interest for astrophysics time before, when he had been Atkinson and Houterman's mentor in their work on thermonuclear reactions in the stars [AH]. Besides, during his last stay in Russia, he had also investigated together with Landau the possible applications of Atkinson and Houterman's model to the evaluation of the internal temperature of stars [GL]. Shortly after he had arrived in the United States, when invited to give a lecture at the Ohio State University, he had chosen to speak about *Nuclear transformations and the Origin of the Chemical Elements* [GG35]. During this conference, Gamow had discussed the possibility that in the inner parts of stars, nuclear processes take place, that give birth to heavy elements starting from light ones. He had considered two kinds of reactions: those induced by protons colliding on nuclei (which are possible because of tunnel effect), and those induced by the absorption of slow neutrons by heavy nuclei (that is, the process studied by Fermi's group).

Gamow's migration from *pure* nuclear physics to its applications to astrophysics, is the result of his desire to move out of a field that, according to him, was being invaded by too many newcomers. He had been the initiator of the quantum interpretation of nuclear processes, but he is now once again bothered by the crowd of physicists working in this field. He feels the need to move towards new frontiers, and these are now represented by astrophysics, where new intriguing questions keep arising. For example: how do heavy elements form out of light ones during the evolution of the universe? What is the origin of the energy emitted by the stars? Some basic ideas had been formulated, but time had now come to make quantitative theoretical predictions.

These were the reasons that induced Gamow and Teller to choose *Stellar energy* as the main theme for the annual GWU conference in 1938. In order not to be unprepared, during the spring of 1938, they work together at a research that will then be published in *Physical Review*: *The Rate of Selective Thermonuclear Reactions* [GT38]. This is a fundamental article, in which stellar energetics is thoroughly investigated.

The starting point of their research is Atkinson and Houterman's 1929 article. Ten years have passed, and nuclear physics is now much better understood. Gamow and Teller consider a general case of a gas consisting of two chemical elements, A and B, with atomic numbers $Z_A$ and $Z_B$, respectively. This gas is compressed and heats up as a consequence of the gravitational contraction of the initial proto-stellar cloud, and nuclei are ionized.

The gas turns thus into a plasma, with ionized (positively charged) atoms at a very high temperature. In this situation, only the Coulomb repulsion between the electric charges $Z_A$ and $Z_B$ of the same sign can work against the direct contact between the ions A and B. This Coulomb barrier can however be penetrated by means of Gamow's quantum tunnel effect, and the ions A and B can thus merge, starting a process of thermonuclear fusion.

Gamow and Teller determine the optimal energetic interval at which this nuclear furnace can work. At the center of the star, the nuclei A and B burn, acting as a kind of fuel, that generates the energy emitted by the star. This energetic interval shows a maximum value, that is called *Gamow's peak*.

Nuclear physics then describes the specific cycles that regulate combustion in various kinds of stellar objects. Gamow and Teller's article is fundamental, as it defines the precise framework within which thermonuclear reactions in the stars can be investigated.

## Hans Bethe Deals with Astrophysics Against His Will

That year, in 1938, Bethe had no intention at all to go to Washington to attend the GWU conference. He had participated in all the conferences during the previous years, but that year he had at first decided not to go to Washington, since he was — by his own admission — "totally uninterested in astrophysics".[2] But Gamow and Teller made him change his mind.

During that conference something big happened, that will make Bethe retrospectively comment: "That turned out to be probably the most important conference that I attended in my life."

It all begins with a student of Gamow's, Charles Critchfield. Short before the beginning of the conference, Critchfield had started investigating the possibility that the energy emitted by the sun is due to a reaction, in which two protons merge to form a *deuteron* (a nucleus consisting of one proton and one neutron), with the emission of a *positron* (a positively charged electron) and a neutrino: $p + p \rightarrow d + e^+ + \nu$. According to his theory, this process is followed by a chain of other reactions, that involve isotopes of hydrogen, helium, lithium, beryllium and boron, and leads finally to the fusion of 4 protons into a nucleus of helium-4, releasing a huge quantity of energy.

However, Critchfield is encountering some technical difficulty in building up his theory. Gamow, although not yet fully convinced that the idea works, makes sure his student and Bethe can meet. Together, the two of them will complete the derivation initiated by Critchfield. The first result is rather disappointing, but simply because at that time the general belief was that the composition of the sun was similar to that of the earth, that is, rich in heavy elements. This implied an inner temperature of the sun of about 40 million kelvin, and therefore a too high emission of energy according to the proton-proton cycle elaborated by Bethe and Critchfield.

---

[2] See Bethe's contribution to the George Gamow Symposium [HPA].

Another positive circumstance occurs at the Washington conference: among the participants that year, is the Swedish astrophysicist Bengt Strömgren. He believes that the sun mainly consists of hydrogen and of a smaller fraction of helium, rather than of heavy elements. Based on his model, the calculated inner temperature of the sun goes down to around 15 million kelvin, and with this evaluation of the solar temperature, the proton-proton cycle can quite realistically describe the amount of energy emitted by the sun.

Back home from Washington, Hans Bethe starts investigating another possible cycle of production of solar energy, the one that involves carbon–nitrogen–oxygen nuclei, and quickly publishes his results.

That same year, in the same months, on the West Coast of the USA, in California, Robert Oppenheimer is about to organize a Symposium on *Nuclear transformations and their astrophysical significance*. Oppenheimer is preparing his lecture about the *physical problem of stellar energy,* that one of his students, George Volkoff, had started working on. Shortly before the beginning of the event, however, Oppenheimer is informed that Bethe, right after the GWU conference, has already published an article exactly on that same topic. Oppenheimer and Volkoff are beaten on the goal line! A brand-new line of research has to be found — and this is a good opportunity for Oppenheimer to show his inventive talent. Together with Volkoff, he starts investigating special cases of stellar evolution, in which gravitational collapse does not reach a state of equilibrium, but continues indefinitely. This work is the prelude to the study of particular kinds of cosmological objects, that the astrophysicist Archibald Wheeler will call *black holes*, years later.

For his results on the energetics of the sun, Bethe will receive the Nobel Prize for physics in 1967. Certainly, one more good reason for him to remember the 1938 GWU conference as a particularly *interesting* one!

## The World Finds Out the Nucleus Can Split

In the meantime, experimental accuracy keeps increasing and new discoveries follow one another, in many fields and in different parts of the world. One of the most important events, both in terms of its scientific value and for the consequences it will have for the entire humanity, takes place towards the end of 1938, at the laboratories of the Kaiser Wilhelm Institute for Chemistry in Berlin. Two chemists, Otto Hahn and Fritz Strassmann, carry out the first experiment of artificial nuclear fission, that is, the process in which a nucleus of uranium, hit by a neutron, splits into two nuclei: one nucleus of barium and one of krypton, whose mass values are much smaller than the uranium mass.[3] A number of free neutrons are emitted in the process, with the release of a huge amount of energy (about 200 MeV).

Hahn and Strassmann's skill as experimentalists mainly consisted in having recognized barium among the products of the split of the uranium nucleus.

Between the end of 1938 and the beginning of 1939, the scientific activities connected to the discovery of nuclear fission and to its theoretical explanation are intense and hectic.[4]

The first big issue was providing the correct theoretical interpretation of the process. An influential role in this context was played by the Austrian physicist Lise Meitner. She had worked together with Otto Hahn in Berlin for a long time, but in July 1938 she had to rapidly leave Germany, due to the Nazi racial laws. From Stockholm, where she had

---

[3] The word *fission* was taken from biology, where it indicates the process of cellular separation.

[4] A detailed description of this phase in the history of physics is provided by Roger Stuewer, in the reference [St], from which we have extracted relevant information.

sought refuge, Lise had an intense correspondence with Otto, who kept her constantly updated about the results of the experimental activity he was carrying out with Strassmann. Otto Hahn benefitted from Lise's expertise on the subject.

In December 1938, when Lise Meitner was confidentially informed by Otto Hahn about the strange results of his experiment, she discusses the results with her nephew, the Austrian physicist Otto Frisch. Together they formulate the explanation of the fission process.

Frisch is in Copenhagen, and he directly informs Niels Bohr about the latest news. Bohr is almost shocked and asks to be informed about all the details, as he is about to sail for the United States, where he will spend the second half of the academic year at the Institute for Advanced Study in Princeton, New Jersey, about 90 km from New York.

Otto Frisch gives Bohr all the information he has about Hahn and Strassmann's shocking results (as well as their possible explanation), on the dock of the harbor in Copenhagen, as Bohr is just about to embark the Swedish-American ocean-liner *Drottningholm*, that will take him and his young colleague Léon Rosenfeld to the United States. Bohr is on his way to Princeton, to work on a theoretical research on quantum electrodynamics, together with Rosenfeld.

It is from this encounter with Frisch on Copenhagen harbor's dock, that Bohr seems to enter a phase of great anxiety. In an urge to understand the physics underlying Hahn and Strassmann's results, he extensively discusses the issue with his colleague Rosenfeld on the ship.

Disembarking in New York, he is personally welcomed at the port by Enrico Fermi and John Wheeler.

Fermi had arrived in New York just a few days before. He had received the 1938 Nobel Prize for Physics in Stockholm, and then embarked the ocean-liner *Franconia*, together with his wife Laura and their two children. Since Laura had Jewish origins, when the fascist regime issued the racial laws, Fermi had decided to leave Italy. His first position in the United States was at Columbia University, in New York.

Wheeler, on the other hand, knew Bohr personally, since he had spent the academic year 1934–1935 in Copenhagen, working in the theoretical group. In 1939, Wheeler was part of Princeton's staff, and on the day of Bohr's arrival in New York he had reached the harbor by train, to welcome his teacher and friend Bohr.

It's Wheeler who, during an interview in 1994,[5] talks about this encounter with Fermi and Bohr at the port of New York, that January 1939. Wheeler reports that, upon arrival, Bohr did not even mention fission, neither with Fermi nor with him. He probably wanted to keep the matter confidential and avoid spreading the news about the experimental discovery (and also its theoretical interpretation), as long as Frisch and Meitner still had not yet published their results.

According to Wheeler, however, Rosenfeld had not been as careful. As soon as they got on the train from New York to Princeton (Bohr had to stay in New York for a couple of days), Rosenfeld told Wheeler everything he knew. Wheeler suggested him to give a talk at the Journal Club, planned in Princeton that same evening at 7:30. And this is how the news about Hahn and Strassmann's experiment started quickly spreading among the American institutions on the East coast.[6]

---

[5] The interviewer is a colleague, physicist Kenneth W. Ford [WF].

[6] Curiously, on a different occasion [Wph], Wheeler himself reports these same facts quite differently: "The act of relaying this particular message [the one about fission, N.A.] was simple: a few words spoken by Otto Frisch to Niels Bohr on the pier in Copenhagen and a

By the end of January 1939, the annual GWU conference is about to begin. The chosen theme is the properties of matter at low temperatures. But the news about nuclear fission causes a sudden change of plans, on the inauguration day. Bohr and Fermi will be the first physicists on stage to communicate Hahn and Strassmann's results to the audience of top physicists that has gathered in Washington.

The news almost certainly caused a twofold reaction. On one hand, it unleashed a purely scientific enthusiasm for the observation of nuclei splitting under the effect of neutron bombardments. On the other hand, however, it must have caused great concern, as the experiment had taken place in Berlin, in Europe, where nationalistic regimes were growing and winds of war were blowing.

One funny anecdote regards Gamow's reaction to the news Bohr was bringing to the world. Bohr had arrived in Washington the afternoon before the beginning of the conference, and immediately reached Gamow's house, to anticipate to him in private the news about nuclear fission. Teller reports having received a phone call from Gamow as soon as Bohr had left, exclaiming "Bohr has gone crazy. He says uranium splits!" [TS]. Considering this reaction, Gamow must have felt somewhat surprised and incredulous. As if he, the "inventor" of the liquid drop nuclear model, could not immediately understand the underlying mechanism.

The following days were both exciting and busy. Physicists were eagerly trying to understand the mechanisms and characteristics of the fission process. The experiment was immediately repeated in Tuve's laboratories at the Carnegie Institute, where it confirmed, in particular, the release of a huge amount of energy. Within short, the experiment was replicated also at the Columbia University (where Fermi was working)

few words spoken to Enrico Fermi and me [John Wheeler, N.A.] by Bohr on the pier in New York."

and at Johns Hopkins. For America, and for the entire world, the atomic age had begun.

> By applying Gamow's liquid drop nuclear model, as explained in Chapter 4, the interpretation of the process of nuclear fission is quite simple and easy to visualize. The uranium nucleus, hit by a neutron, deforms, and elongates along an axis, assuming the typical shape of a rugby ball. Two forces play against each other: the nuclear attraction at a short range and the Coulomb repulsion between the protons. If the deformation continues, two lobes progressively form in opposite directions (imagine the shape of a peanut), until they completely separate. This is how nuclear fission takes place: the initial heavy nucleus of uranium splits into two nuclei, both with a mass that is approximately half the mass of the original nucleus — plus a number of neutrons.

It is according to this model that Lise Meitner and Otto Frisch managed to provide an interpretation of the fission process [MF]. Niels Bohr reached a similar conclusion too, that he exposed in a quite complex article he wrote together with Archibald Wheeler [BW]. We will go back to this further on in our book

During Weiner's interview, Gamow seems to be very disappointed: how come he never thought about the possible implications of his own model? A bitter feeling, that will accompany Gamow in the years to come.

> One more short note about fission. Most probably, processes of nuclear fission had already been produced by Enrico Fermi and colleagues in 1934, in a series of experiments during which uranium composites were bombarded with slow neutrons, to induce nuclear transmutations. Fermi was looking for transuranic elements. Remarkably, the German chemist and physicist Ida Noddack, in an article she sent to Fermi, explicitly put forward

the suggestion that, most likely, nuclear fission processes had been generated in Rome. Emilio Segré, one of Fermi's most famous Italian colleagues, in one of his works wonders why their group had not considered this possibility, during the hectic activities they were carrying out between 1934 and 1935. Many years later, he directly asked Fermi, who replied that back then the values for the nuclear binding energies were still so uncertain, that, to him, fission did not seem to be possible.

## A First Step Towards Cosmology

Despite the extraordinary news about nuclear fission that had shocked the world, towards the end of the 1930s Gamow keeps investigating the field that is giving him the greatest personal satisfaction — that is, the applications of nuclear physics to astrophysics.

It is probably during one of his morning phone-calls to Teller, that Gamow suggests that they should start working together on the formation and evolution of galaxies (or rather, *nebulae*, as they used to call them). Gamow is here about to take one more courageous step — the one towards cosmology, an old passion of his, since his university years in Petrograd. The path he is about to start walking is filled with problems, mainly because there was too little data available from cosmological observation. Gamow comments his first works with Teller on this subject as follows: "You see, at this time the age of the universe was wrong, and therefore our conclusions were wrong" [AIP] — something absolutely obvious at the time Gamow and Teller published their results in *Physical Review* [GT39], that is at the beginning of 1939. Also, they were lacking a description of the cosmos that gave a physical content to the very first cosmological models.

Anyhow, despite the flaws therein, Gamow was emotionally bound to this article, as it was the last paper he ever wrote with Teller during the years they spent together at the GWU. In 1941, Teller will move to

Chicago to work with Fermi, and the following year to Berkeley, to join Oppenheimer. During the celebrations organized at the GWU for Teller's 60th birthday, Gamow will pay him a personal tribute, remembering that particular work about nebulae, that they had written together many years earlier.

Meanwhile, Gamow keeps getting more and more involved in astrophysics. His research extends to various aspects of stellar evolution, analyzing the different morphology of stars, from white dwarfs to red giants. Together with his student Critchfield, he starts investigating in detail the internal parts of the stars, where nuclear fusion reactions take place. Yet, the climax of this research will be reached after the trip to Brazil that Gamow takes, together with his wife, during the summer of 1939.

## Neutrinos in Rio de Janeiro

As already mentioned, Gleb Wataghin, who at that time was in San Paolo, Brazil, sends Gamow an invitation to give a series of lectures there. The visit to Brazil that Wataghin and colleagues organize for Gamow also includes a stop-off at the *Casino da Urca*, a famous elegant resort, located in a wealthy residential neighborhood, offering a hotel, a theater and, of course, a gaming hall. On that occasion, Gamow meets for the first time the young researcher Mario Schönberg. Most probably, the promoter of this encounter had been Wataghin himself.

Gamow is very positively impressed by Schönberg, and suggests him to spend a year at GWU, to work with him on nuclear astrophysics. For this, Schönberg will be granted a Guggenheim scholarship.

In November 1940, they publish their first synthetic work together, in *Physical Review*. The title sounds still somewhat hesitating: *The possible role of neutrinos in stellar evolution* [GS40], but the idea is extraordinary and absolutely original.

Gamow and Schönberg make the hypothesis that, in the inner parts of a star, under the effect of its gravitational collapse, the density and temperature of the electrons increase and cause the following reactions:

$$e^- + N(A,Z) \rightarrow N(A,Z\text{-}1) + \nu_e, \; N(A,Z\text{-}1) \rightarrow N(A,Z) + e^- + \bar{\nu}_e$$

In the first reaction, a nucleus with A nucleons and Z protons captures an electron, transforming a proton into a neutron, with the emission of a neutrino. In the second one, the newly formed unstable nucleus undergoes a $\beta$ decay, that brings it back to the original nucleus, emitting one electron and one antineutrino. The described cycle obviously also applies to the case where the initial nucleus is hydrogen — that is, one proton — and what decays is a neutron.

The immediate consequence of these two consecutive reactions is the production of a huge number of neutrino/antineutrino pairs. These particles interact very feebly with matter, and, unless the density of matter is extremely high, they escape from the star, carrying a huge quantity of energy with them.

The process suggested by Gamow and Schönberg represents an important cooling mechanism for many stellar objects, such as white dwarfs and neutron stars. In the case of supernovae, this process contributes to trigger the explosion of the star itself. The massive emission of neutrinos from the center of the star, in fact, reduces its internal pressure — consequently, its external parts start collapsing gravitationally.

This idea is further developed by the authors in a following publication, with which they lay the foundations of the theory of supernovae: *Neutrino theory of stellar collapse* [GS41]. According to their theory, the emission of a flow of neutrinos anticipates the enormous luminosity of the supernova's explosion, which is what makes it visible. It is April 1941.

These neutrinos, that rapidly emerge from the supernova and then get lost in the sidereal space, remind Gamow and Schönberg of the rapidity with which the gamblers' money vanishes into the cash registers of the casino. So why not name the physical process as *URCA* process after the name of the Casino? Gamow expects the reviewer of the article to ask the authors for an explanation of the strange name given to the physical process, and gets ready for an answer. He jokingly coins the idea that URCA is an acronym that stands for *UnRecordable Cooling Agent!* Luckily, they will never be asked to justify the name.

Gamow's and Schönberg's theory highlighted the fundamental role of neutrinos in the dynamics of supernovae explosions. Further investigations about processes involving neutrinos in this kind of explosions, were carried out by various other scientists too. The theory was finally confirmed experimentally in 1987, when laboratories on Earth measured a flow of neutrinos 2–3 hours before the visible observation of the SN 1987A supernova.

---------- **10** ----------

# If I'd Got 500 Degrees Kelvin, It Would Have Been Unpleasant

*Gamow: I don't like the word "big bang"; I never call it "big bang"[...].*

*Weiner: That's interesting because the names are practically synonymous in the popular mind: Big bang equals Gamow. You know that?*

*Gamow: Yes. Because you see, I did it — with Herman and Alpher — 20 years ago, and then [...] I calculated what the temperature is; I got 5 degrees [kelvin], I got 7 degrees, then I got 7, then I got 5 — just to see what it would be, because if it would be 500, it would be unpleasant. Outside would be boiling water.* [AIP]

## A New Idea of Cosmos

When his young student Ralph Alpher sees George Gamow step into his office, waving the latest issue of *The Soviet Journal of Physics*, he immediately understands something big is going on. In fact, Gamow is about to inform him that the research he has been working on for his doctoral thesis, that is, the growth of condensations in a relativistic homogeneous and isotropic expanding medium, must be abandoned. Evgenii Lifshitz has just published his results about exactly the same subject in the above-mentioned journal.

Alpher had been working for quite some time on this topic, and must have felt really disappointed by the news. Besides, those were quite hard times for him, having to split his time between the GWU, where he attended evening classes, and his job at the Applied Physics Laboratory at Johns Hopkins University.

Gamow immediately assigns a different topic to his PhD student. It is something he has been personally mulling over since quite some time: the formation of chemical elements in the universe and in particular, the possibility that they form in its early phases.

It is 1946, and the topic is definitely not a new one. It is one of those moments when *things hang in the air*. The scientific community has accepted the idea that the evolution of stars is determined by thermonuclear reactions taking place in their inner parts, in a continuous contrast with the strong gravitational field generated by their own mass. Such reactions mainly depend on the physical effects that Gamow has predicted in his works. He had, in fact, already faced this problem and during a conference he had given at the Ohio University back in 1935, he had illustrated how a sequence of thermonuclear transformations inside the stars can describe quite well the mechanisms of production of a wide range of chemical elements.

Between the end of the 1930s and the early 1940s an important question arises about the distribution of abundances of chemical elements in nature. The point is figuring out whether the observed abundances of chemical elements can comply with the hypothesis that they are produced in the inner parts of stars, or if they might have formed by means of thermonuclear reactions much earlier, during a phase of the evolution of the universe previous to the formation of stars (the one that is usually called *pre-stellar phase*).  One reason to support the latter idea, i.e., that of a cosmological origin of chemical elements, consists in the fact that their distribution in the universe appears to be uniform. In 1942, Subrahmanyan Chandrasekhar and Louis Heinrich had investigated this idea. However, they reached the conclusion that the cosmological origin of elements cannot be explained by a single model in a situation of thermal equilibrium.

Chandrasekhar and Heinrich present their results at the 1942 GWU conference, focusing on *Stellar evolution and cosmology*. Based on their results, Gamow works on a new cosmological model, that will remain as a milestone in the evolution of cosmology. Gamow's hypothesis is that in the early phases of evolution of the universe, *dynamic* processes of formation took place — in complete contrast with the idea of a *static* universe in thermal equilibrium. He announces his idea during a conference at the Washington Academy of Sciences. In building it up, Gamow is largely inspired by Lemaître, who had put forward a similar idea between 1927 and 1931.

Georges Lemaître is a fundamental figure in the history of cosmology. Born in Charleroi, Belgium, in 1894, he initially studied engineering, then physics and mathematics. He also had a religious vocation, and was ordained priest in 1923, after three years in seminary.

Lemaître was mainly interested in relativistic cosmology and astronomy, and he studied in top academic environments, such as the University of Cambridge (where he first meets Arthur Eddington) and the Massachusetts Institute of Technology in Boston. During his stay at the MIT, he has the opportunity to receive first-hand astronomic information: Edwin Hubble is measuring the distance of nebulae, by means of the Cepheid method, and Vesto Slipher is calculating their recession velocity.

These observations will allow Lemaître to formulate a relation of proportionality between the recession velocities of the galaxies and their relative distances.

Lemaître did not know at that time that Alexander Friedmann (the Russian physicist who had been Gamow's professor in Leningrad) had already published in 1922 an article about the general solution of Einstein's relativistic equations in *Zeitschrift für Physik*. When Lemaître starts working on that problem, he reaches Friedmann's same conclusions: an expanding universe is a natural solution of the relativistic equations. The idea that the universe is expanding dynamically, instead of being stationary, is taking shape.

It is quite surprising that Lemaître did not know about Friedmann's article. After all, Friedmann's paper had sparked quite some interest in the scientific community. It had also been criticized by Einstein himself (even though he had later on withdrawn his criticism). Lemaître could probably not read German.

It might be due to these linguistic problems, that history never fully recognized Lemaître all the merits he deserved. We here refer in particular to the physical law that establishes a proportionality between the recession velocity of galaxies and their relative distance. In a publication in 1927, Lemaître had reported the formula he had calculated, providing even an indication as to the numeric value of the proportionality constant. Yet, his work had been published only in French in the *Annales de la Société Scientifique de Bruxelles*, and thus went rather unnoticed. Eddington, who had received a copy directly from Lemaître, at first did not pay any attention to it. Later on, in 1930, realizing the importance of the result presented therein, he quoted it in one of his own articles, and made sure it was published in English, on the Acts of the Royal Astronomical Society. However, the proportionality relation went lost in the translation. Due to this, the fundamental cosmological property usually goes under the name of *Hubble's law*, based on the results Edwin Hubble published in 1929, two years after Lemaître's note.

In 1931, Lemaître publishes (this time directly in English) in *Monthly Notices of the Royal Astronomical Society*, a compendium on his line of reasoning about an expanding universe. He tries to describe the origins of cosmos and does so in quite a literary style, while trying to give shape to a possible physical model. In this journey backwards in time, Lemaître introduces the idea of a *primordial atom*, that breaks up and gives birth to all cosmological structures.

It is precisely on this vague idea of a primordial high-density system suggested by Lemaître, that Gamow's first model of an expanding universe is based. He substitutes Lemaître's primordial atom with a *superdense* state, consisting of nuclear matter. This initial state then evolves, through processes of neutron capture and nuclear fission.

For Gamow, this is just the beginning of a phase of production of new ideas, that will then take a more concrete scientific shape between 1946 and 1948.

## The Strange Story of the αβγ Article

Gamow's fundamental article about his new line of research is dated September 1946. Its title clarifies the topic: *Expanding Universe and the Origin of Elements* [GG46]. Gamow suggests the idea that, if the density of the primordial universe is high, Friedmann-Lemaître's cosmological model implies that the cosmos expands very rapidly. And for this reason, in order to calculate the distribution of the abundances of chemical elements, the idea of a static universe is not realistic — what the model requires, is a dynamic situation in continuous evolution.

In the light of what we know today, Gamow's model is not correct in many aspects. It predicts a primordial state of high density and high temperature neutrons, that progressively cools due to its expansion. Neutrons coagulate into complex structures, giving birth to the various atomic species, by means of β decays. According to this model, the abundance of hydrogen we measure today in the cosmos, is due to a fraction of neutrons that underwent β decays before coagulating into nuclear structures. Despite being imprecise, the model represented a first important step that finally gave physical content to Lemaître's hypotheses.

The *annus mirabilis* is 1948. Gamow clarifies his ideas on some specific points. The initial state of the universe consists of a heavily compressed neutron gas, therefore extremely hot. These neutrons start decaying into protons, electrons and neutrinos (β decay), as the pressure of the gas diminishes, as an effect of the cosmologic expansion. In the gas, that now contains both neutrons and protons, single neutron-proton pairs fuse together to create a deuteron, releasing electromagnetic energy: $n + p \rightarrow d + \gamma$. Going back to his 1946 article, according to Gamow this process triggers a number of nuclear reactions, based on a simple mechanism: by capturing a neutron from the gas, a nucleus undergoes a transition to a nucleus with the mass number increased by one unit. Then, if necessary, in order to gain stability, inside the nucleus a neutron transforms into a proton, via β decay. A new chemical element is thus formed — as it contains one more proton compared to the initial one. That is the successive element on Mendeleev's Periodic Table of chemical elements. Gamow at this point asked himself another question: can this mechanism describe effectively the cosmological formation of all chemical elements, with abundances fitting the observed values?

In order to answer, it was necessary to calculate the probability with which nuclei can capture neutrons, and how fast this process is. This was the very difficult task Gamow entrusted his student Alpher with. In this, Alpher had the chance to use the results of the investigations carried out in Los Alamos during the building of the atomic bomb, which had been disclosed at the end the Second World War.

Alpher and Gamow publish their first results in a short note sent to *Physical Review* [GG48a]. This work will be remembered among cosmologists as the $\alpha\beta\gamma$ *article*. Gamow had here set up one of his memorable jokes.

> Maybe aware of the fact that this article was going to represent the beginning of a new era for cosmology, Gamow wanted the initials of the authors' names to recall the beginning of the Greek alphabet. Alpher's name provided the $\alpha$, and obviously his own name stood for the $\gamma$ — but what about the missing $\beta$? He needed someone whose name began with a B. So, Gamow thought of his good friend Bethe. He did not involve Bethe in the writing of the article — actually, he did not even inform him — but simply added his name to the list of authors, and sent it to the journal. Near Bethe's name, he prudentially wrote *in absentia*. Although quite unbelievable, the story is true and ends with the editor of *Physical Review* removing *in absentia*, and with Bethe's great appreciation — from then onwards, Bethe included the $\alpha\beta\gamma$ article in his own curriculum vitae.

As mentioned above, the article by Alpher, Bethe and Gamow will prove to be incorrect for many reasons, but still, it represented the framework of a model for the cosmological nucleosynthesis of light elements (deuterium, $He^3$, $He^4$, $Li^7$) that, after further elaborations by other physicists, fits observational data quite accurately. In this sense, their theory can be considered as one of the cornerstones of the model of evolution of the universe from an initial hot and dense phase — the one that, some years later, will be called the *Big Bang model*, and usually associated to Gamow's name (even though he never called it with this name...).

Investigations carried out after the $\alpha\beta\gamma$ article will show that the cosmological nucleosynthesis in the primordial universe cannot produce elements beyond lithium — the main reason being the absence of stable nuclei with mass numbers 5 and 8. Heavier nuclei are produced in the stars.

## The Cosmic Background Radiation Appears for the First Time

In spring 1948, when the $\alpha\beta\gamma$ article is just published, Gamow is already pointing towards an even more ambitious goal. He wants to approach a new phase in the study of the model of a dynamic evolution of the universe, that is, the processes of formation of cosmological structures.

Gamow sums up and publishes his conclusions in a short article for *Physical Review* in June that same year [GG48b]. This paper gives him the opportunity to provide numerical estimations of some of the fundamental parameters of the model. He points out that the results he has obtained in his two latest works (the 1946 and the $\alpha\beta\gamma$ articles) "permit us to get certain information concerning the densities and temperatures which must have existed in the universe during the early stages of its expansion."

The steps of his line of reasoning are rather simple. Primordial nucleosynthesis begins with the formation of deuterons, in a plasma strongly dominated by electromagnetic radiation (photons). A deuteron, formed by the fusion of a neutron and a proton, is immediately split by the many high energy photons in the plasma, if the temperature does not drop below a value of $10^9$ K (a billion kelvin), that corresponds to approximately 0.1 MeV (one hundred thousand eV).[1] It takes a few minutes for

---

[1] The energy associated to the thermal motion of the particles of a given physical system is proportional to the so-called absolute temperature T, i.e. to the temperature counted from the absolute zero point, which (rounding off) corresponds to –273 degrees Celsius. Absolute temperatures are given in *kelvin*. As an example, in a system of particles at temperature $10^4$ K (ten thousand kelvin) the average energy of a single particle is about 1 eV.

the temperature of the primordial universe, Gamow explains, to drop from the extremely high initial temperatures down to $10^9$ K. Gamow also makes an evaluation of the density of nucleons at this temperature, and estimates it around $10^{18}$ (one billion billions) per cubic centimeter.

Gamow's model implies that the electromagnetic radiation, that permeates the cosmos, has the same characteristics of an electromagnetic radiation inside a cavity, in thermal equilibrium with the walls of the cavity itself — that is, the so-called *black-body radiation*. This radiation, described in function of the various wavelengths (from radio waves to gamma rays), is represented by a typical bell-shaped curve — the famous *Planck distribution*, that gave birth of quantum physics.

Since during the cosmological expansion the density of the radiation energy (which is the dominant energy component in the primordial universe) decreases faster than the density of matter, there must have been a cosmological epoch during which these two densities were comparable. Gamow estimates this happened at a temperature of approximately $10^3$ K, when the universe was about 380,000 years old.

From that moment onwards, matter takes over and dominates radiation. The plasma can start condensing into cosmological structures, as an effect of gravity. In the meantime, the photons that constitute the electromagnetic radiation, stop interacting with matter, and their life continues independently — maintaining a memory of what matter looked like at the time of their decoupling.

As the temperature of the universe keeps dropping during its expansion, these photons, that no longer interact with matter, keep pervading the cosmos. Their black-body bell-shaped curve is preserved, and its peak progressively moves towards longer wavelengths.

From Gamow's estimation of the density of nucleons at the temperature of nucleosynthesis, $10^9$ K, it can easily be derived that the cosmos should currently be pervaded by a black-body radiation of about 5 kelvin. At this temperature, the peak of the electromagnetic radiation curve can be found on wavelengths of the order of a centimeter, that is the wavelength range of the microwaves.

The existence of such a microwave background radiation is not explicitly named by Gamow, but is a direct consequence of his model.

> The fact that light nuclei should have formed during such a short period of time, just a few minutes, compared to the cosmological scale, led to consider the theory as quite bizarre. In April 1948, when Alpher has completed his research and is about to defend his PhD dissertation, the GWU informs the news bulletin for former students, and the news start circulating among journalists. A famous cartoonist publishes on the Washington Post a comic vignette, in which a man (who actually resembles a bomb) is reading a newspaper, with a big title: "Scientist says world was created in five minutes", and wonders: "Five minutes, eh?"

## Some Disagreements with His Collaborators

Robert Herman, a student of the Applied Physics Laboratory at Johns Hopkins University, starts collaborating with Alpher, who had so far been accurately investigating the details of the physical processes of primordial nucleosynthesis. Herman had arrived at Johns Hopkins some years before Alpher, but had so far been working in the field of applied physics. Their scientific collaboration immediately works very well, and in September 1948 they write their first article together [AH48a], shortly after followed by another paper, written with Gamow [GG48c].

During the summer of 1948, something curious happens — a first sign that Gamow is taking distance from his collaborators. Gamow is in Los Alamos, on a temporary leave from the GWU. While there, pursuing his investigations on the formation of cosmological structures, he writes a note to *Nature* [GG48d], and sends a copy also to Alpher and Herman. The two read Gamow's manuscript in detail, and find a couple of mistakes

in the evaluations about the transition phase from the radiation dominated regime to the matter dominated one. Alpher and Herman quickly
inform Gamow, who, instead of stopping the publication of his article,
asks them to rapidly write a note and send it directly to *Nature*. Gamow
also informs the journal about the upcoming note by his collaborators.

Following Gamow's advice, Alpher and Herman send their note to
*Nature* [AH48b], listing the mistakes encountered in his evaluations and
revising them. Quite surprisingly, they conclude the note with the following acknowledgments: "Our thanks are due to Dr. G. Gamow for the
proposal of the topic and his constant encouragement during the process
of error-hunting."

The story could end here, if the note didn't also contain a comment
in which Alpher and Herman mention that according to *their* evaluations,
the current temperature of the cosmic background radiation is around
5 kelvin. This note will become the argument with which Alpher and
Herman will claim their own *exclusive* priority in predicting the cosmic
background radiation — a subject which they both will further develop
in the years to follow.

> Big steps forward in the description of the primordial universe
> were made by many physicists, when it became clear that the
> initial plasma, originally described by Gamow as a neutron gas,
> should actually consist of all the particles (or their elementary
> constituents) whose presence is compatible with the temperature
> of the plasma. So, for instance, at the time when nucleosynthesis
> begins, the plasma should consist of neutrons, protons, electrons,
> positrons, neutrinos and anti-neutrinos (kept in balance by
> weak interaction), as well as photons. This aspect was first made
> evident in the early 1950s by the Japanese physicist Chushiro
> Hayashi [Ha], and then used by Alpher and Herman together
> with James Follin Jr. [AFH]. In this way, the primordial plasma
> (that Alpher named *ylem, material from which elements were*

*formed* [AIP]) enriched itself with all the constitutive elements that particle physics was progressively discovering.

This continuous development in particle physics affected also Alpher and Herman's evaluations of the temperature of the background radiation, and inevitably their estimations had to be continuously adjusted, within a realistic range [AH88]. As Gamow humorously put it during his interview with Weiner, "the important point is that the theory of a hot and dense primordial universe does not imply a present cosmos so hot to make water around us boil!"

The most incredible aspect of this story, however, is what did NOT happen for about fifteen years after the articles that Gamow alone or with his collaborators had written in 1948 and shortly after. Despite the many new and extraordinary predictions, it is really surprising that the community of astrophysicists never activated in order to try to reveal the existence of a background radiation, to be looked for in the microwave region of the spectrum. It was as if no one had understood the real relevance of Gamow's innovative ideas for cosmology. Even Alpher and Herman, after their publication with Follin in 1953, stopped researching on this and left the Johns Hopkins Laboratory to move to two different industries.

## An Annoying Disturbance in the Radio Wave Antenna

It was only in 1964 that two radio-astronomers, Arno Penzias and Robert Wilson, ran into a signal due to the cosmic background radiation. They were studying the radio waves emitted by our galaxy, by means of an antenna at the Bell Telephone Laboratory in New Jersey. They perceived a continuous *disturbance*, a radio "noise" with a wavelength of 7.5 cm, equivalent to a temperature of approximately 3.5 kelvin. This curious story can be found also in Nobel Prize Steve Weinberg's popular book, *The first three minutes* [Wei].

Trying to figure out where that strange background noise came from, Penzias and Wilson made several attempts to eliminate it, even removing some birds' nests from the antenna. A group of cosmologists in Princeton came to help, providing a possible cosmological interpretation: more than a disturbance, what Penzias and Wilson had captured was a precise signal, due to a cosmological background radiation. The two groups, the experimental and the theoretical one, agreed to publish their results in two combined articles in the same issue of the *Astrophysical Journal* [PW, DP].

Quite *curiously*, the theoretical article quotes Gamow, Alpher and Herman only for their results on cosmological nucleosynthesis, but overlooks their prediction of a background radiation. Obviously, this, together with some other *lapses of memory,* caused Gamow great disappointment. In 1967, he wrote an article together with Alpher and Herman in which they re-established the correct historical sequence of events [GG67].

Gamow's premature death did not allow the scientific community to fully recognize him his true merits while he has still living.

Alpher and Herman, became with time very bitter about what had happened between them and Gamow. They even argued that Gamow's sometimes excessive superficiality had harmed both himself and, as a reflection, them, and that their work had often been discredited because of this [AH88].

The physicist who most of all contributed to recognize to Gamow and colleagues their priority in the development of the cosmological model that predicts both primordial nucleosynthesis and the cosmological background radiation, was Steve Weinberg, with the before-mentioned book.

After Penzias and Wilson, other radio-astronomers measured the background radiation, at a wavelength slightly shorter than 7.5 cm. In order to confirm the prediction, more accurate measurements were needed, and for this, a suitable measuring instrument had to be sent in space, because short wavelengths are absorbed by the atmosphere. This was finally achieved in 1989, when the satellite COBE (COsmic Background Explorer) was put into orbit. It was equipped with the radiometer FIRAS, which measured the spectrum of the background radiation on a wide range of wavelengths. This spectrum perfectly fit the typical bell-shaped curve of a black-body, at a temperature of 2.735 kelvin [COBE]. When these results were presented at a conference of the American Astronomical Society in January 1990, they were welcomed with a standing ovation. More recent measurements of the temperature of the cosmic radiation give 2.725 kelvin.

The success of Gamow's model caused the definitive end of the stationary model of the universe, strongly supported at the end of the 1940s by Fred Hoyle, Hermann Bondi and Thomas Gold (from the University of Cambridge), as well as by a number of other renowned cosmologists. For a long time, a harsh controversy went on between the supporters of a hot expanding universe and those defending a static cosmological model. In 1949, Fred Hoyle, who was also a great communicator, during a famous program on BBC, trying to make fun of the cosmological model starting from a very hot primordial phase, called it the *Big Bang* model — something that also distorted the core concept of the model. However, despite the fact that the name had been given ironically, with the success of the theory, *Big Bang* became its official name. This name contributed also to the great popularity of the theory.

Gamow never spared Hoyle his sharp remarks. Playing with the well-known bitter rivalry in Cambridge between Hoyle and Martin Ryle (the radio astronomer whose measures on galaxies over large scales seemed to contradict the stationary model), Gamow asked his second wife, Barbara, to write a short poem, recalling this debate. During the interview with Weiner, Gamow insists with Barbara that she reads some verses. We here choose just a couple of strophes — that we consider particularly meaningful and amusing[2]:

"Your years of toil,"
Said Ryle to Hoyle,

---

[2] The complete text can be found in https://history.aip.org/history/exhibits/cosmology/ideas/ryle-vs-hoyle.htm.

"Are wasted years, believe me.
The steady state
Is out of date.
Unless my eyes deceive me,
My telescope
Has dashed you hope."

Said Hoyle, "You quote
Lemaître, I note,
And Gamow. Well, forget them!
That errant gang
And their Big Bang."

Cartoon by Eli Dwek

# It Was Very Amusing to Create Codes and Ties

*Gamow: Well, I didn't know what to do. When I saw the Watson-Crick model of DNA, then I knew what to do.*

*Weiner: Well, tell me about that. That's what I had hoped to learn from you. How did that come about?*

*Gamow: Well, to be exact, because I remember very well this day, I was for some reason visiting Berkeley and I was walking through the corridor in Radiation Lab, and there was Luis Álvarez going with Nature in his hand and he said, "Look, what a wonderful article Watson and Crick have written." [...]*

*Weiner: Had you been following these developments beforehand?*

*Gamow: No.*

*Weiner: You knew about DNA, though?*

*Gamow: No, I didn't. I don't think so. [AIP]*

## Deciphering the Code: a New Mathematical Enigma

Towards the end of the 1940s, a phase of Gamow's career was coming to an end. He had provided crucial and unique scientific contributions, for which he was going to be remembered as the greatest pioneer of Big

Bang physics. It had not been an easy path. The atmosphere within his research group appears to be sometimes tense and arguments with some other colleagues unavoidable.

Time had now come for Gamow to embark into a new adventure.

The opportunity came in 1953, during a visit he paid at the Radiation Laboratory of the University of California, in Berkeley. That specific laboratory had carried out research for the Manhattan Project during the Second World War, and was now developing a new particle accelerator. By means of this particular accelerator, the great experimental physicist Luis Álvarez will discover new particles, and for this receive a Nobel Prize in 1968.

Rather unexpectedly, Gamow's new adventure, the one that he refers to in his autobiography as *an extravagant deviation* [GG70], begins right here, in Berkeley. In a corridor of the Radiation Laboratory, he comes across Luis Álvarez, who is reading an article that has just been published in *Nature*. It's the work of two renowned scientists, Francis Crick (a physicist and crystallographer) and James Watson (biologist), about the structure of the DNA molecule [CW].

This paper reports the results of their recent discovery, mainly the fact that the hereditary genetic information of all living beings is contained inside the DNA molecule, which is structured as a sequence of four different nitrogenous bases. These were great news and not only within biology. Gamow's reaction to the article was so enthusiastic, to decide to personally engage in the research — venturing into a field that was very far from his own area of expertise.

During Charles Weiner's interview, to the question as to what had induced him to dedicate himself to genetic research, Gamow replies by quoting Ehrenfest, who had once described the birth of a new interest with the following words: "a spider that sits in a corner in a big web, and he waits; then some fly or something gets caught; he quickly goes there. So, I am just sitting and waiting, listening, and if something exciting comes, I just jump in" [AIP]. Gamow feels like that spider, whose attention is suddenly caught by something new and unexpected. Logically speaking, Gamow's deviation into biology is quite weird; yet, it can be understood remembering that this is the scientist who throughout his entire life loved to be a pioneer, to be the first one to walk the uncrowded paths of new fields of exploration.

But what was the aspect that caught Gamow's attention when it came to genetic transmission?

It all starts with one of those mathematical tricks that he loves so much. When reading Watson and Crick's article, besides finding it extremely interesting, Gamow realizes that some of the basic processes illustrated therein are not yet fully understood. One of the main issues is the correct description of the mechanisms according to which the genetic information contained in the DNA is translated, during a process that ends with the synthesis of proteins. Based on the fact that proteins, in the human species, are a combination of 20 amino acids, and knowing that during the process of translation a ribosome inside the cell (a protein-based structure with the task of building proteins) picks the amino acids according to the sequence of the 4 bases of the DNA, Gamow has a brilliant idea. It's all about how "one can get 20 out of 4", and that can be done by "counting the number of all possible triplets formed out of four different entities" [GG70]. By so doing, Gamow makes a completely arbitrary, yet correct, hypothesis that the 4 bases come together in *triplets* (that is, groups of 3). Starting from this assumption, it is not difficult to calculate the number of "groups of 3" elements one can get out of 4: four triplets made of the same kind of base, four triplets made of all different bases and twelve triplets made of 2 similar and 1 different base. Summing up: 4 + 4 + 12 = 20.

Mathematically, one can actually get 64 different combinations, but Gamow made the assumption that the order of the bases within a triplet is irrelevant. Thus, counting out all the triplets made of the same combination of bases (despite the order in which they appear), he managed to reduce the number of combinations from 64 to just 20.

Gamow once again had the right intuition, this redundancy in the genetic code is actually something true. However, in biology this property is due to another aspect: the reason is not that the order of the bases in a triplet is irrelevant, but rather that different triplets code for the same amino acid. This happens for triplets that differ for just the base in third position. Gamow ingeniously indicated the correct directions, but disregarded aspects he was unfamiliar with.

His great intuitions consisted, in synthesis, in assuming that there exists a correspondence between triplets of bases and amino acids and in suggesting a mechanism by which the 64 combinations reduce to just 20. Both ideas were proved ideally right, even though they needed to be revised, since they are based on chemical (not mathematical) properties.

Believing to have found *the* solution, Gamow gets in contact with Crick and Watson to submit them his theory. Since this solution is quite bizarre, at first they are quite skeptical. However, getting to know Gamow better, amazed by his enthusiastic approach to research and by his incredible stubbornness, they decide to formally involve him in their research. With Crick in particular, Gamow establishes a true and long-lasting friendship [Cri].

Gamow sums up his ideas in a short letter to *Nature* [GG54] first, and then in a more complete paper that he sends to the *Proceedings of the National Academy of Sciences,* of which he is a member. Unfortunately, the feedback he receives is not a positive one, probably because of his somehow too superficial approach. After all, a renowned physicist who, all of a sudden, starts writing scientific articles on biology could hardly gain credibility. To prevent a possible rejection of the paper, Merle Tuve asks him to withdraw the article. Gamow does so, but simply readdresses it elsewhere, to the Royal Danish Academy of Copenhagen, in order to have it published in their *proceedings* [GG55]. This time, he is careful enough to at least remove Mr. Tompkins' name from the list of authors — something that probably had helped discredit his work.

The *gamovian* theories about the transmission of the genetic code are too hypothetical and, as already said, will be proved wrong. Facing such a problem only by means of mathematical speculations was too risky and practically impossible. However, Gamow is stubbornly determined to solve this *enigma*, as he states towards the end of his autobiography [GG70]. He involves other physicists (such as Edward Teller) in the research, as well as experts in decoding encrypted languages working for the army, whom he knows thanks to his collaboration with Navy. He even starts an exclusive club — the *RNA Tie Club* — a "haphazard collection of Gamow's friends" as Crick calls it. The members of the club are engaged in a role play where every person corresponds to one amino acid, identified by a particular tie and tie-pin, with the chemical symbol on (tie and tie-pin were specifically tailored). In Gamow's intentions, all this was aimed at creating as many occasions for gathering as possible, since he is convinced that sharing and discussing ideas are the essential premises for scientific progress.

The mechanism of the genetic code will be deciphered much later, at the beginning of the 1960s, by three biologists: Marshall Nirenberg, Har Gobind Khorana and Robert Holley. For their discovery, they will win the Nobel Prize for Medicine in 1968. The genetic information is transferred according to a mechanism that is not at all as elegant as Gamow had hoped and foreseen with his theoretical and purely mathematical model. With his typical sense of humor, about this Gamow will say that the solution to the problem "has the indisputable advantage of being correct, elegant or inelegant" [GG70].

Gamow's major contribution to the genetic code problem mainly consisted in the role he covered for all the scientists he was able to gather together. He was an enthusiastic and inspiring force for all of them. Crick himself defined Gamow's presence in the group as fundamental during that particular phase. He introduced biologists to a mathematical method that, although only partially effective, at least was not burdened by all those elucubrations that typically slow down biological research.

When his interviewer asks Gamow which one among all his contributions had had the greatest impact on the scientific environment, his answer is quite surprising. "I don't know", he replies, "probably the code". Acknowledging that his contributions to nuclear physics actually were the ones that allowed him to get to know, among others, Niels Bohr, and thus opened up the path for his future career, still the most exciting, or even the most amusing of them all was biology.

## Longing for His Native Land

One personal issue probably strongly influenced Gamow's evaluation of his own role within the field of biology. Thanks to his contribution to genetics he managed, at least partially, to reconnect with his motherland. After his definitive defection from the Soviet Union and having become an American citizen, he was considered as a *persona non grata*, especially among Russian physicists and astronomers. This is something Gamow painfully mentions in his own autobiography. He had no contact at all with the Soviet Union, and was probably hesitant to establish any, fearing a possible reaction of the regime on his soviet colleagues. Ageing, this detachment from his motherland must have become very difficult for him to bear. Finally, he found an opening, thanks to Nikolai Viktorovich Luchnik (1922–1993), a biophysicist who was responsible for the Institute of Biophysics of the Academy of Sciences of Moscow. He is the author of a book whose title is reported by Gamow to Weiner as "Why I resemble my daddy" — a summary of all the most recent discoveries in the field of genetics. In his book, he writes as follows: "The alphabet of the genetic code was deciphered because many different people had given thought to it: geneticists, biochemists, crystallographers, mathematicians and many others in England, the U.S.S.R., France, U.S.A., Germany, Japan, and other countries. The deciphering of the code of living molecules would probably have taken a different direction had it not been for an unusual scientist, living in a tiny hamlet, Boulder, Colorado, U.S.A. His little house on Sixth Street doesn't look like the neighboring houses. The crooked chimney on the shingled roof, the carved shutters, the apple tree under the window, all give the feeling of an old-fashioned dacha in a Moscow suburb. But this is not surprising, because the owner's name is Gyorgy Antonovich" [AIP].

Gamow proudly shows these few lines to Weiner during his interview, giving the feeling that finally, at the end of his career, he had somehow received a recognition from *his* native land. Actually, this might explain why his "deviation" into biology was so important for Gamow. It ended up representing the reason why he was going to be appreciated and remembered also in the Soviet Union.  Perhaps this is a romantic interpretation and certainly Gamow's multifaced personality is very difficult to frame within just one context. He was an extraordinary scientist, unique in his approach to research. Gamow had a natural sense for sudden intuitions that helped him figure out unexpected solutions and indicated alternative routes to follow whenever facing a problem. He had a wide culture that he combined with an innate enthusiasm for everything that was *new,* enriched with a very special sense of humor. In his mature days, one more element completes the picture of Gamow, as he reveals his sensitive side, especially when he looks back with nostalgic eyes at his life, at that *line of Universe* that he wasn't given a chance to complete.

## 12

# A Morning Train to Princeton

*"So, every other Friday I took a morning train to Princeton, carrying a briefcase tightly packed with confidential and secret Navy projects."*

*"A number of years ago an article in a national magazine described my contribution to the development of the hydrogen bomb as that of bringing Edward Teller to this country."*

*(from the autobiography)* [GG70]

### Nuclear Physics Loses Its Innocence

Gamow unexpectedly dedicates only a very short part of his autobiography to the activities he carried out in the United States. The section dedicated to the years he spent in America is quite schematic and divided in three parts: research and academic activities, science popularization and consultancies for military organizations. The present chapter of our book deals with the last part.

Between the 1930s and the 1950s, many physicists, both experimental and theoretical ones, were involved in the war projects of the countries they belonged to. Considering Gamow's rather problematic position as a Russian immigrant in the USA, it is particularly interesting to explore his approach to military issues, as well as the way he related to this reality.

Gamow had in his youth been involved with the army back in Russia, when he was an undergraduate student in need of a job, and had been promoted to the rank of colonel, in order to hold a teaching position at the Artillery School, and get a salary. His involvement with the army was therefore almost accidental, and he often expressed quite ironic comments about his clumsy attempts to fit into the military context.

On a broader level, there is no clear evidence as to Gamow's political orientation, especially in his mature days. In his younger years in Russia, he had taken part to students' protests, the before mentioned Young Turks; but later on, when he lived in America and the Second World War was about to break out, he does not seem to have expressed a definite political orientation, unlike many other physicists of that time. One may wonder whether this was due to caution, as he somehow still felt like a foreigner, or if it was lack of interest in political issues whatsoever.

Between the end of 1938 and the beginning of 1939, just a few years after Gamow's arrival in the United States, the scientific community, and physicists above all, started being involved in the delicate ongoing political events. Otto Hahn and Fritz Strassmann had shown that nuclear disintegration can be induced by bombarding uranium atoms with neutrons. In this process, every nucleus of uranium splits into two lighter nuclei, that share (even though unevenly) most of the mass of the original nucleus. Two features of this fission process drew physicists' attention.

First of all, during the process much more energy is released compared to what is normally produced during other kinds of nuclear transitions.

Moreover, free neutrons are released that, moving within a material that contains uranium, can in turn trigger further fission transitions and, under particular circumstances, originate a chain reaction.

One of the first physicists to predict the possibility of a chain reaction was Hungarian-American Leo Szilard, whose role was to become fundamental in the development of nuclear energy.

As a direct consequence of Hahn and Strassmann's experiment in 1938, the scientific community realized that nuclear fission, if controlled, could be used both to produce energy and to build bombs with a devastating exploding power. A thorough investigation of all the details of nuclear fission processes began, trying to identify what kinds of atomic nuclei split when hit by neutrons — the so-called *fissile* nuclei. Uranium was identified as a good fissile material because of the presence of its isotope uranium-235. Natural Uranium, however, consists of just 0.7% uranium-235, whereas 99.3% is uranium-238 that is not fissile. A second fissile element was also identified: plutonium-239, which can be produced by hitting uranium-238 with neutrons.

In February 1939, Meitner and Frisch had provided a first interpretation of nuclear fission in terms of the liquid drop nuclear model. Shortly after, in June that same year, Niels Bohr (at that time a visitor at the Institute for Advanced Study in Princeton) and John Wheeler provided an in-depth investigation of the phenomenon. [BW]

> It is worth observing here that none of the above-mentioned theoretical works mentions the fact that the idea of modelling the nucleus as a liquid drop originally belonged to George Gamow, who had first presented it in London on February 7, 1929 during a meeting of the Royal Society. In connection to this, it is just amazing that in 1994 John Wheeler, during an interview with his colleague Kenneth Ford, says: "I've been told that the liquid drop model of the nucleus appeared as one of many ideas in an early edition of Gamow's book on nuclear physics, but I don't remember ever having looked that up myself. I know that Bohr had taken up the liquid drop model ... from the fall of 1935 onward." [WF]

The discovery of nuclear fission put an end to a period of extraordinary development for nuclear physics, that had begun with Becquerel's discovery of radioactivity and continued with Rutherford's experiments, in times when research was carried out in a spirit of pure scientific interest. A period that has appropriately been denominated by science historian Roger Stuewer as the *age of innocence of nuclear physics* [St].

With the discovery of fission, the situation completely changes: new incredible scenarios suddenly open up — during a historical phase characterized by upcoming dramatic events on an international scale. Nuclear physics, and physicists in first person, all of a sudden are involved also in ethical, political and economic issues.

## Physicists Activate — And Secrecy is Imposed

Within the new political framework that is taking shape from 1939, all political authorities, or at least those of the main powers involved in the upcoming war, are urged to make a decision whether and how to eventually apply the new potentials offered by nuclear physics. The evaluation of the pros and cons of such a decision is made extremely difficult by the awareness that the production of nuclear weapons requires a very advanced level of technology and logistics and, as a direct implication, huge investments of both human and financial resources.[1]

Germany only planned the development of nuclear reactors. Many factors influenced this decision, the main one being that Germany believed they would win the war very quickly.

In the United States, as earlier said, the scientific community on one hand immediately activated in order to thoroughly investigate the mechanisms of fission; on the other hand, an ethical reflection on the possible implications of applying fission to military purposes began. The physicists that most strongly felt the urge to act were former European immigrants, probably more sensitive than others to the political climate on the other side of the Atlantic sea.

Among them were three Hungarian physicists: Edward Teller, Eugene Wigner and the before-mentioned Leo Szilard. They had a clear sense

---

[1] A detailed and richly documented description of these aspects, concerning the historical period from 1939 until the development of the Manhattan Project, can be found in [Well]. This reference is our main guideline in the present chapter.

of how potentially dangerous the situation was and decided to inform President Franklin Delano Roosevelt. Albert Einstein was the most distinguished among all physicists working in the US at that time, and was asked to sign a letter to the President, to make him aware about the situation. Einstein, Szilard and Teller wrote the letter together. They provided accurate details about the meaning of the discovery of fission and described all its possible implications and risks. They had written a shorter and a longer version of the letter, but decided to send to the President the longer one. And this certainly was the right choice.

A crucial role in the process of the preparation of Einstein's letter was played by Alexander Sachs, an influent economist and banker in New York. He was in contact with many physicists, and was very impressed by the scientific results about fission. He immediately understood their possible dramatic consequences. Sachs had a "ready access to the White House", and used this opportunity to personally call the attention of the President on the need for the US administration to act promptly. It was at the end of a crucial meeting with Sachs, that Roosevelt finally concluded: "Alex, what you are after is to see that the Nazis don't blow us up" [Hew].

Shortly after, the Advisory Committee on Uranium was established, within the National Bureau of Standards. President Roosevelt made his first political moves very rapidly, always pressed by Alexander Sachs, who persuaded the President to purchase all necessary materials to build the first nuclear reactor. Despite this rapid beginning, however, until 1941 all the steps towards the definition of an organic program were quite slow and uncertain, mainly due to the huge investments required.

In the United Kingdom, investigations around the possibility of using fission for war purposes started, mainly thanks to the contributions of Otto Frisch and Rudolph Peierls. They were both Jewish and had managed to flee Nazi Germany. A Committee (called MAUD) was constituted, aimed at exploring military applications of nuclear fission. However, the UK could not stand alone for the technological and economic investments required to develop this kind of project. Therefore, during summer 1941, a delegate of the MAUD committee was sent to the United States, to get in contact with American physicists and urge them to start a common plan of development of nuclear projects.

> From the early days of the *nuclear age*, a sense of awareness rose among physicists first, and within short among politicians too. It was clear that from that moment onwards, talking about new results in the field of nuclear physics required great caution. During spring 1940, a Reference Committee was founded within the National Research Council of the United States, with the specific task to control the policy underlying all scientific publications that touched upon topics of military interest. A sub-committee was also established, specifically for fission. Previous to their publication, all editors of scientific journals had to submit to the Reference Committee copies of all articles that required caution, and wait for approval. This procedure was applied until June 1945.

## PO Box 1663, Santa Fe, New Mexico

A significant turn in the decision of the United States to undertake a program to study the feasibility of a nuclear bomb occurred between the end of 1941 and the summer of 1942. Vannevar Bush, the director of the Office of Scientific Research and Development (OSRD), founded in June 1941 by President Roosevelt, played a crucial role in this decision.

Bush was moved by the fear that Germany might be working at a nuclear program, and convinced President Roosevelt that the USA needed to urgently plan the realization of an atomic bomb. In the summer of 1942, a specific organism was constituted, the Manhattan Engineer

District, under the aegis of the historical US Army Corps of Engineers. The crucial center for the scientific coordination of the new agency was the Columbia University in New York; together with the University of Chicago and the University of Berkeley, these were the three main institutions involved in the realization of nuclear weapons.

The first fundamental requirement in order to trigger a chain fission reaction was the availability of fissile material. Two were the possibilities. The first one consisted in enriching natural uranium in its fissile component uranium-235, compared to uranium-238 (more abundant, but not fissile). The second one, consisted in the production of plutonium-239, obtained by bombarding uranium with neutrons.

The largest plant for the enrichment of uranium was built in Oak Ridge, Tennessee. In order to produce plutonium, nuclear reactors were built, according to a prototype realized by Enrico Fermi's team at the University of Chicago, the famous reactor named Chicago Pile-1. With this reactor, the first controlled chain reaction ever was achieved. At the same time, the construction of a large-scale reactor was planned, in Hanford, Washington State.

Along this complex preliminary phase of scientific and technological development, towards the end of 1942 the Manhattan Project was being shaped. Although the pre-existing plants for both uranium enrichment and plutonium production were to be maintained, time had come to establish a new top-secret laboratory where to gather all available top experts in physics, chemistry, material sciences and engineering, to work at the realization of the first atomic bomb.

Brigadier General Leslie Groves was nominated as Director of the Manhattan Project. He in turn chose J. Robert Oppenheimer as leading scientist in the program. Together, the officer and the scientist evaluated the best location where to build the new research center: the Los Alamos Laboratory was born. Its address was the famous PO Box 1663, Santa Fe, New Mexico. During all the years of the Project, all correspondence and materials addressed to the Laboratory passed through this office, the only reference point for all that happened on the site.

The bombs dropped in August 1945 on Hiroshima and Nagasaki were physically produced here.

## Chatting with Einstein

Gamow didn't take part in the Manhattan Project. Even though he had formally become an American citizen already in 1940, he had to wait until 1948 to get the clearance from the American administration to partake in programs of nuclear applications for military purposes and to access nuclear sites covered by military secrecy. Suspected of being ideologically close to communism, due to his past in the Red Army, Gamow, the theoretical physicist who had provided fundamental contributions to nuclear physics with his works on the tunnel effect, the liquid drop model and the *Gamow factor*, was kept out of America's most important research project ever.

> Albert Einstein did not take part in the Manhattan Project either. Retrospectively, the official justification for his exclusion was his being an exquisitely theoretical physicist, scarcely inclined to practical applications; for this reason, he was not considered as a *suitable* scientist. However, the real reason was most probably his pacifistic leanings, something that influenced very negatively his involvement in the project.

Having been excluded from the Project does not mean that Gamow had no connection whatsoever with military activities. Certainly hurt by the establishment's mistrust and denial to access all secret nuclear sites, yet he "very happily" accepts an offer to work part time as a consultant for *conventional* military issues for the Bureau of Ordnance of the Navy Department. Gamow's task is to study the propagation of shock and detonation waves caused by conventional explosives — all subjects under the jurisdiction of the *Division of High Explosives.* According to the conditions in his contract, he must spend two afternoons a week in an office located in the buildings owned by the Navy, in Constitution Avenue.

Based on this, the picture of Gamow that we can draw is that of a well rooted person on quite a small territory: he moves between the George Washington University, the Navy Department — both located in the institutional part of Washington — and his home, in Bethesda district, just a few miles away.

What Gamow remembers with most pleasure about his collaboration with the Bureau of Ordnance, are the frequent meetings with Albert Einstein, with whom he established a very close relationship. Albert Einstein was a consultant for the same *Division of High Explosives,* but since he stayed in Princeton, it was difficult for him to reach Washington and participate to meetings or personally check how research was progressing. So, Gamow offered to be the connection between him and the *Bureau.*

Gamow tells about his journeys by train from Washington to Princeton every other Friday to meet with Einstein at his home (and also adds accurate descriptions of the pullovers he used to wear). Gamow and Einstein dedicated their mornings together to discuss the various projects the Navy submitted to their expert opinion. But at lunchtime and during the afternoon they could finally immerse into discussions about the subjects they both were most passionate about: astrophysics and cosmology. Unfortunately, Gamow does not provide details about the content of their conversations. Even if he had investigated what Einstein was working on, he always kept this subject confidential. However — he confessed — he couldn't retain from occasionally casting a glance to the many papers scattered all over Einstein's desk. They were filled with formulae of tensorial analysis — a mathematical formalism used to investigate unified fields — a research subject Einstein dedicated his last years to.

The regular meetings with Einstein represented for Gamow an incredible occasion to deepen a both professional and personal relationship with one of the greatest figures of his time. Gamow will throughout his life remember these encounters as "unforgettable".

Despite not being personally involved, Gamow could certainly figure out that somewhere in the United States (probably in New Mexico...) something big and top secret was being planned. He realized it had to be something of primary importance when he was personally affected by a sudden decision of the Navy. He tells the anecdote in his autobiography.

Gamow had had a brilliant idea (double-checked and approved by Einstein too) as to how to optimize the power of detonation waves by means of a rather complex mathematical model. He submitted his idea to the Navy, that decided to immediately start testing it. The first chosen site for the tests was the Navy base at the Naval Support Facilities in Indian Head (Maryland), just about 40 km south of Washington, along the Potomac river. Unfortunately, this site turned out to be unsuitable for the range of the experiment Gamow had in mind. But since the Navy was really interested in this project, they contacted an important producer of explosives, in order to plan further testing. All of a sudden, however, the highest ranks of the Army informed Gamow that his project had been downgraded from the top to the bottom on their list of priorities. The complete absence of explanations for such an abrupt decision, made Gamow realize that the Army was now focusing on something bigger, with top priority.

Gamow had quite a clear idea what route needed to be followed in order to start applying fission to the development of nuclear weapons. What was not so clear to him, as well as to all other scientists not personally involved in the Manhattan Project, was the complex technology that was being developed in Los Alamos and in all connected laboratories.

All strategic activities were top secret. And because of the impossibility to understand what was going on, all over the Country a feeling of uncertainty and discomfort was spreading.

## A Red Stamp: Atomic Bomb

The first among the scientists involved in the Manhattan Project to perceive that widespread feeling of unease, was Henry Smyth, at that time professor and Chair at the Department of Physics at Princeton University. He had personally covered important roles within the Project and had been a member of the Uranium Committee in 1941. Between 1943 and 1945, while he is a teacher in Princeton, he also works as a consultant and is the associate director of the Chicago Metallurgical Laboratory. He commutes on a weekly basis between these two institutes.

Smyth realizes that Americans need to be at least partially informed about the ongoing research. With this aim, he discusses the opportunity to write an official document with the scientists involved in the Manhattan Project and with General Groves himself. This document should contain the main scientific goals of the Manhattan Project, respecting all secrecy restrictions. When he receives approval in 1944, he is ready to start working at the report. Groves has granted Smyth access to all available material about the Manhattan Project, including top secret files. Within short his office at the Palmer Institute in Princeton turns into a kind of bunker, with anti-theft systems and armed guards outside the door. Smyth's document, in the author's intention, is meant to provide an explanation of the physics behind the atomic bomb, and at the same time also set the rules as to what could be made public and what needed to remain top secret.

Smyth works actively at his report and submits several draft copies for approval to the main figures responsible of the Manhattan Project, including Oppenheimer. The British scientists involved were particularly concerned about spreading this kind of information. Their initial scepticism, however, eventually faded away.

The final approval of Smyth's document [SR] is signed by President Harry Truman on August 9, 1945, once the Project's senior managers have given their green light. The report is officially released according to a precise procedure. Copies of the report will first be provided to the radio media on August 11, 1945 and then to the newspapers on the following day.[2]

> The document is issued during the very last phases of the Second World War. On July 16, 1945 the first nuclear explosion had been tested in Alamogordo, in the Jornada de los Muertos desert, New Mexico. It has showed the efficacy of the plutonium based nuclear bomb. During the Potsdam Conference, that took place between July 17 and August 2, 1945, an ultimatum was issued to Japan. Upon Japan's refusal to surrender, on August 6 the first atomic bomb was dropped on Hiroshima, followed on August 9 by a second one on Nagasaki.
>
> Three days later, Smyth's document was released.

It was the first time a document of this kind was issued. It had gone through many problematic phases before being officially released, including the formulation of a suitable title. In the first version, in fact, the title simply read *"Atomic Bombs"*, but it was followed by a very long

---

[2] A detailed description of the birth of the Smyth Report can be found in "Smyth Report", https://en.wikipedia.org/wiki/Smyth_Report, accessed April 6, 2021.

and bizarre subtitle: "*A General Account of the Development of Methods of Using Atomic Energy for Military Purposes under the Auspices of the United States Government, 1940–1945*". This title was obviously almost impossible to use, and was soon changed to a much simpler *"Atomic Energy for Military Purposes"*. However, the document is most well known as the *"Smyth Report"*.

The Smyth Report is definitely worth reading.[3] The first part is dedicated to the explanation of the physics behind the phenomenon of fission: it describes all available fissile materials and illustrates practical problems, such as uranium enrichment processes and plutonium production. In the second part, a description of the various sites in which the project was carried out is provided, with drawings and photographic documentation. Of course, for security reasons very much is missing. Still, the amount of sensitive information made public during that delicate historical phase is quite astonishing. No wonder that the choice to issue the Smyth Report was severely criticized in the 1950s by President Dwight Eisenhower.

General Groves wrote the *Foreword* to the Report, while Smyth himself wrote its *Preface*: their explanations as to the purposes of the document are very different.

Smyth states in the *Preface* that "The ultimate responsibility for our nation's policy rests on its citizens and they discharge their responsibilities wisely only if they are informed. The average citizen cannot be expected to understand clearly how an atomic bomb is constructed or how it works but there is in this country a substantial group of engineers and scientific men who can understand such things and who can explain the potentialities of atomic bombs to their fellow citizens. The present report is written for this professional group." In Smyth's vision, therefore, the document is addressed to a small group of scientists, to provide them with all the

---

[3] Text available on https://www.osti.gov/opennet/manhattan-project-history/publications/smyth_report.pdf.

instructions in order to reach out to common citizens and explain the goals of the secret operations carried out in their Country.

In Groves' *Foreword*, on the contrary, the tone — a military one — is very different. He made a clear distinction between what can be made public and what cannot, specifying that potential violators of the instructions would be "subject to severe penalties under the Espionage Act."

## Dear Dr. Bush — Where the Boundary Should Be Placed?

As already mentioned, George Gamow had to wait until 1948 before he was granted authorization to access nuclear sites and top-secret files. From the point of view of his scientific career, Gamow had left behind nuclear physics already back in 1938, when he decided to venture on the paths of astrophysics and cosmology.

In the summer of 1939, while the physical community was frantically digging out the properties of fission, Gamow left for a holiday at Copacabana beach, near Rio de Janeiro, together with his wife. This was the journey during which he got inspired to name the effects of neutrinos in stellar evolution as URCA process. Shortly after, he dedicated himself to another field at the frontier of astrophysics, the investigation of Red Giant Stars [GG39b].

In spite of this apparently complete change of interests, Gamow probably continued investigating the nucleus on his own, and had most probably made realistic guesses about what his colleagues in the Manhattan Project were working at.

Early in August 1945, after the two Americans bombs were dropped on Japan, Gamow wondered whether it was time to start talking more openly about the physics that lays behind nuclear weapons. He plans to write a book, aimed at providing explanations about the fundamental principles of nuclear physics and its role in both stellar evolution and in the building of nuclear reactors and weapons.

Gamow feels pressed to write this book, but at the same time is very cautious; he wants to comply with all the restrictions imposed by national security regulations. His approach is very straightforward. On August 12 — just three days after the nuclear bombardment on Nagasaki — he addresses a letter directly to Vannevar Bush. His question is: how much can be revealed about nuclear physics and its applications to military issues?

It is a handwritten letter, on two pages. The English spelling is at times incorrect, as is quite typical of Gamow, but the formulation of his arguments and questions is extremely accurate and clear. He underlines that, despite not being involved at any level in the project of the atomic bomb, his expertise has allowed him to get his own insight about the possible developments in that field. And as he is about to give a class in nuclear physics and also wants to write a book on the same subject, Gamow asks: what am I authorized to say and where do I have to stop? Where goes the limit between what is permitted and what is not?[4]

Vannevar Bush replies within just two days: his advice is to refer to the Smyth Report — that had been released on the same day Gamow sent his letter — and to stick to the indications provided within.

Gamow writes his book in an astonishingly short time. The preface to the first edition, published by Macmillan Company in New York, is ready in September 1945. Gamow states that he has chosen the topics for his book according to the official instructions received from the War Department of the United States Government. And then jokingly explains how he intends to make complicated subjects understandable to his readers: "There are also a number of pictures which the author drew simply as a relaxation during the process of writing the book; it is hoped that these pictures will also give some relaxation to those who read it."

---

[4] This is reported and discussed by Alex Wellerstein in http://blog.nuclearsecrecy.com/2013/01/18/george-gamow-and-the-atomic-bomb/.

This book, whose title is "Atomic energy in cosmic and human life" [GG47], became soon a best seller. We will come again to this book later on, when discussing George Gamow's extraordinary activity in popularizing science.

Other authors who wrote about nuclear physics in the postwar period were concerned about carefully sticking to the government's security rules too. An interesting example is a volume that goes simply under the title *Nuclear Physics*; it is a collection of notes taken during Enrico Fermi's lessons at the University of Chicago in the first term of 1949 by three of his students, namely Jay Orear, A. H. Rosenfeld and R. A. Schluter. Far from being aimed at a wide public, it is an advanced text for physics students, summing up Fermi's expertise and extraordinary didactic skills. We mention it here because the section "Theory of Chain Reactions" begins with the following statement: "For security reasons Dr. Fermi followed fairly closely the article given below. To avoid clearance delays, we reprint it directly, by courtesy of Science." The chapter then reports the presentation Fermi gave at the Chicago meeting of the American Physical Society in June 1946 — everything complies with General Groves' instructions.

## Post-war Army Consultancies

In July 1946 the United States were running nuclear tests at the Marshall Islands, in the Pacific Ocean, within the so-called Joint Army/Navy Task Force One project. Gamow was invited to assist to the tests. He must study the effects of the shock waves produced by nuclear explosions on the military ships placed around the Bikini atoll. Gamow commented very briefly that the experience had been "interesting and stimulating".

During the following years, Gamow's activity as a consultant for the Army continued in a new direction, at the same time challenging and amusing: he started collaborating with the *Operation Research Office,* run by Johns Hopkins University, once again just a few blocks from home, in Connecticut Avenue where the District of Columbia enters into Maryland for about one mile.

The study consists in the application of operational analysis techniques to solve problems of military logistics. Gamow was dealing with complex simulations of war games, by means of a newly invented computational method, the so-called *Monte Carlo Method.* This method had been invented by Stanislaw Ulam, Enrico Fermi and John von Neumann within the Manhattan Project. Gamow now tries to apply the Monte Carlo Method to optimize military attack strategies, during field battles involving tanks.

The Method will later be applied to many different kinds of problems, whenever a solution cannot be obtained by means of traditional analytical methods. Nowadays the Monte Carlo method is used in various different fields, such as engineering, business sciences, medical physics and computational chemistry.

Finally, Gamow was also a consultant for a number of companies, in research projects in the field of atomic weapons, in connection with U.S. military services.

He also collaborated with *Convair,* a company that was working at the realization of the first intercontinental ballistic missile [Har2].

## Finally, the Clearance

The American administration's doubts as to involve Gamow in building nuclear weapons definitively fell in 1948, when he was finally granted a clearance and, together with it, received a formal invitation to spend a year at the Los Alamos Laboratories. Free from suspicion and mistrust,

Gamow officially accepts the invitation in a postcard, addressed to Norris Bradbury, director of the laboratory. A reproduction of this postcard can be found in Gamow's autobiography, witnessing how much that invitation to Los Alamos meant to him.

He takes a one year leave from GWU and moves to Los Alamos, to partake in the development of a hydrogen based thermonuclear bomb, better known as H-bomb. In this case the physical process involved is the nuclear fusion of two isotopes of hydrogen (namely deuterium and tritium).

Gamow is very excited for this opportunity: the research is interesting and, moreover, he can finally reconnect with Edward Teller. Their long-lasting collaboration had been abruptly interrupted when Teller had left the GWU to join the Manhattan Project.

In a letter to Gleb Wataghin, dated November 1947,[5] Gamow expresses his desire to go back to Brazil, for a conference tour and for some tourism in Rio de Janeiro. In that letter, he mentions the fact that Teller would also be interested in going to Rio and directly suggest the three of them meet there the following year, to *do something fun* together. Gamow and Teller's friendship was still strong, despite the distance.

However, upon arrival in Los Alamos in 1949, Gamow faces quite a difficult situation. Norris Bradbury had taken Robert Oppenheimer's role as director in October 1945, and was dealing with a completely different situation, compared to that of the previous years. Many of the physicists whose roles had been determinant in the development of the fission bomb had left Los Alamos at the end of the war to move back to their original institutes.

---

[5] Private letter kept by Vladimir Wataghin, Gleb Vasilievich's son; we thank V. Wataghin for allowing us to use the quote.

In the meantime, the new project for the fusion bomb had begun — and physicists were split, due to opposite scientific orientations: the new director Bradbury on one side and Teller on the other. When Gamow reached Los Alamos, the subject of the research had moved from purely scientific topics to engineering — something very far from Gamow's taste and aspirations.

Besides, the world was undergoing dramatic political changes: during the war, the building of the atomic bomb was aimed at endowing the US with a weapon to fight the advance of the Nazis. Now, in a new political framework characterized by the Cold War with the USSR, all projects in Los Alamos had to be developed in a completely different perspective.

Perhaps due to all these reasons together, Gamow did not contribute with any relevant work to the research in Los Alamos. About this period, in his autobiography Gamow mainly focuses on his fear losing his friendship with Teller, who had become a completely different person after the time in Los Alamos.

Teller, however, will always remember Gamow as one of his best friends ever.[6]

## Starring as a Lecturer in the Washington District

Towards the end of his stay at GWU, Gamow was contacted by his old-time colleague Maurice Shapiro, in order to give a course on advanced physics to the scientists of the Nucleonic Division at the Naval Research Laboratory. At the time of the Manhattan Project, Shapiro was leading a research group studying the effects of underwater nuclear explosions and from 1949 he coordinated the studies on cosmic rays at the Naval Research Laboratory.

---

[6] See the message Edward Teller sent to the George Gamow Symposium in 1996 and reported in the records of the conference [HPA].

Well aware of his friend Geo's talent as a lecturer, Shapiro asked him to give weekly lectures at the Naval Research Laboratory. These lectures were initially aimed exclusively at Navy members, but soon became events of great interest for the entire scientific community in the area of Washington. High rank officers attended Gamow's conferences, wishing to be updated about advanced scientific subjects directly from such a famous physicist.

They say people arrived early to the weekly appointments with Gamow who, with his lively, captivating and at times funny manners, always managed to thrill the audience talking about particle physics, astrophysics and nuclear physics. As reported by Eamon Harper [Har2], Maurice Shapiro describes him as follows: "we could always be sure of learning something new and exciting. (....) The breezy and humorous tone in which he recounted these developments was reminiscent of the bright and expository style that made his popular books so readable" [Sh].

# About Mr. Tompkins and Other Stories

George Gamow ends his autobiography *My World Line* with these words:

> "*Do I enjoy writing books on popular science? Yes, I do.*
> *Do I consider it my major vocation? No, I do not.*
> *My major interest is to attack and to solve the problems of nature, be they physical, astronomical or biological. But to 'get going' in scientific research one needs an inspiration, an idea. And good and exciting ideas do not occur every day. When I do not have any ideas to work on, I write a book; when some fruitful idea for scientific pursuit comes, writing lags.*" [GG70]

Gamow writes about science along his entire career, from the mid-1930s until the end of the 1960s. Writing is his passion, something he loves doing in parallel to his activity as a researcher. He writes popular books, aimed at a wide public of lay readers, people who are potentially interested in science, but in need of a simple and clear language to understand its complexity. Being an experienced lecturer, expert on the subjects he writes about, and combining his expertise with his natural talent as a communicator and a good lot of fantasy, Gamow can undoubtedly be considered as one of the greatest writers of scientific popular books of the past century.

In his autobiography, he personally explains how he began writing popular books. It happened almost spontaneously, he says, when he realized that he had a natural skill in making complicated concepts understandable. As he explains, "probably it is because I love to see things in a clear and simple way, trying to simplify them for myself, I learned how to do so for others" [GG70].

Gamow had actually started writing when he was a young student in Leningrad. Back then, he used to write scientific articles for two famous Russian newspapers, *Istvestia* and *Pravda*. In particular, a column on the lower part on the second or third page of the *Pravda*, was dedicated to science, and that was where Gamow wrote about stars, cosmic rays and much more. That extra job, which he was paid for, was very rewarding, especially because he realized he was really talented.

There are many different reasons why Gamow decided to dedicate himself to writing books in his mature days — both professional and private ones. On one hand, whenever his research rhythms slowed down or he was lacking new interesting ideas to investigate, writing filled up his otherwise rather empty days. However, most probably, as Karl Hufbauer points out in the biographic monograph of the National Academy of Sciences dedicated to Gamow [Hu], writing books meant also something else to him. In the extremely conventional academic *milieus* Gamow used to attend, he stood out as a unique figure, difficult to understand and to frame within just one context. Probably due to this, his work was often underestimated, and sometimes not even taken seriously. His qualities as a scientist were seldom fully recognized, "stained" as they were by his desecrating and too unconventional behavior. Getting older, this must have given rise to a sense of bitterness, that was somehow compensated by the success of his popular books. Feeling free to share with enthusiasm the most recent scientific discoveries with a constantly wider audience, probably brought him back a sense of fulfillment, that he otherwise was missing.

Besides, his activity as a writer represented an extra source of income, something he definitely was in need of, especially in the 1950s.

Hufbauer also points out that from a personal point of view, writing allowed him to carve out a private space, in which he could distract himself from the problems connected to the end of his relationship with his wife Rho and to the move from Washington to Boulder. It had been a difficult, yet necessary, decision, in order to keep physical distance from extremely painful private memories [Hu].

Whatever the reason for Gamow's passion for writing was, what's sure is that he left a series of very interesting and pleasant popular books behind him. His full production is certainly impressive for the total number of volumes, but most of all for the wide range of topics he wrote about. He stretched from scientific fiction (quite different from what commonly goes under the denomination of *science-fiction*), to the popularization of scientific subjects and to biographies of prominent scientists.

Covering Gamow's production thoroughly would lead us far beyond the scope of this book. Yet, we'll try to outline the main thematic areas of his writings and comment upon the most relevant and popular books.

## Meet Mr. Tompkins

The first series of books authored by Gamow is represented by the stories centered on a protagonist: Mr. C. G. H. Tompkins. He is the hero of a series of simple, yet very impactful, popular books. In making up this character, Gamow borrows the surname from the young researcher who had been his personal assistant during the summer of 1935, in Ann Arbor. The combination of letters that form his name sound amusing to him. As to the initials of Tompkins name, Gamow takes them directly from physics: C stands for the velocity of light, G for the gravitational constant and H for Planck constant. Tompkins is not a physicist, but a shy and curious bank clerk, who all of a sudden gets passionate about physics. He has strange dreams about imaginary worlds. In these dreams, he sees the effects of some physical properties that normally can be perceived only in the microscopic world. By using the trick of the dreams, Gamow manages to explain concepts of relativity and quantum physics, that otherwise would be too difficult to express in a simple way. The adventures of Mr. Tompkins begin in 1937, with the first short story entitled *Toy Universe*. At first rejected by a couple of magazines (*Harper's Magazine* and *The Atlantic Monthly*), it was finally published in *Discovery* the following year, in 1938. The story has an immediate success, and is the first of a series of stories with Mr. Tompkins as the protagonist. His many adventures are collected in one volume, *Mr. Tompkins in Wonderland* [GG39], whose title recalls the famous book by Lewis Carroll. Gamow somehow compares his protagonist to Alice, who lands, one after the other, in surrealistic worlds, where incomprehensible things happen.

Probably Alice was a character that really intrigued Gamow. So much that in a paper he wrote for *Nature* on the idea of a "negative proton" — an idea which, by the way, turned out to be wrong — he quotes an excerpt from *Through the Looking Glass* (the sequel of *Alice in Wonderland*):

> Alice laughed. 'There's no use trying,' she said. 'One can't believe impossible things'. 'I daresay you haven't had much practice,' said the Queen. 'When I was your age, I always did it for half-an-hour a day. Why, sometimes I've believed as many as six impossible things before breakfast.'

It is quite significant that in a scientific paper in which the existence of a new particle is about to be suggested, Gamow refers to *believing at least six impossible things* — Alice's life recipe, that very much reminds of the spirit with which Gamow used to carry out research.

Despite the fact that he was a made-up character, Mr. Tompkins slowly became Gamow's adventure mate. The books where Mr. Tompkins is the protagonist, in fact, focus on the same topics Gamow is personally researching on. Somehow, Tompkins seems to be walking in Gamow's own footsteps. And on one occasion, he even treated him as if he was a real person. It is the case of an article published in *Physical Review* in 1939, *On the Origins of Great Nebulae* [GG39a], in which Gamow, in the acknowledgments, thanks both Hans Bethe for the precious discussions and C. G. H. Tompkins for "having suggested the topic."

Mr. Tompkins appears again in 1944, in *Mr. Tompkins explores the atom* [GG44], a book about atomic physics, and in 1953 and 1967 when Gamow has "deviated" into biology and is investigating the frontier fields of genetics, in *Mr. Thompkins learns the facts of life* [GG53], and *Mr. Tompkins inside himself: adventure in the new biology* [GG67a].

The books about Mr. Tompkins were reprinted several times, in different editions. They were translated into many languages and can be found collected in one volume, *Mr. Tompkins in Paperback* [GG65]. The preface to this edition was written by 2020 Nobel prize winner Roger Penrose. More recently, in the 1990s, a British physicist, Russell Stannard, published a more modern version of the work, in which all the examples are updated to the developments of nowadays physics (*The new world of Mr. Tompkins* [RS]).

## Astrophysics, Nuclear Physics, Mathematics, Biographies: Books for All Tastes

Gamow's production of popular books is much wider than just the books about Mr. Tompkins adventures. His works stretch from nuclear physics, to cosmology, astrophysics and biology. He combines his scientific expertise with his personal passion for literature, arts, music, languages and travelling.

A peculiar aspect of Gamow's literary production is that he wrote all his books in English, which was not his mother tongue. His written English is correct and very far from the bizarre English he used to speak.

His friends used to say that he spoke *gamovian* — a kind of mixture of Russian, German, English with a great number of terms he picked in the many countries he had visited.

Gamow was also particularly skilled at drawing. Many of his books are enriched with his own sketches and comics, that give them quite an unusual artistic touch.

As regards the subjects Gamow wrote about, it is possible to group his production of popular books according to the topics, in different areas.

First of all, the astrophysical one. In this area, he touched upon subjects of great interest for his public, especially the American readers, who were eager to understand all about space. Between the 1950s and the 1960s, in fact, America is beginning to "conquer space", with the first launches of the Apollo missions, that will reach the moon in 1969. Among the titles we here name *The Birth and Death of the Sun* [GG40] in 1940, reprinted in 1964 with a new title, *A Star called the Sun* [GG64]; *The Moon* [GG53a] in 1953; a *Planet called Earth* [GG63] in 1963.

Nuclear physics is a topic to which Gamow dedicates great space. The most important book is *Atomic Energy in Cosmic and Human Life* [GG47], that we mentioned earlier, when discussing the rules America's government had set in 1945 as to treating themes connected to the Manhattan Project.

This book is very rich in details and accurate in the presentation of scientific contents, and at the same time very fluid and easy to read. The properties of nuclear fission, absolutely new to the wider public, are very clearly presented. He starts with an overview of the chemical processes — or rather, *alchemical* as Gamow ironically calls them. Gamow illustrates the dramatic potentials of nuclear weapons — often inappropriately called *atomic* — as compared to traditional explosives (such as trinitrotoluene, for example). In the mid part of the book, Gamow clarifies that nuclear physics, besides being a carrier of tragedies for humanity, is also responsible for fascinating things, such as stars shining and energy production.

The book had an immediate success. In June 1946 Gamow wrote a new preface for the second edition, this time for Cambridge University Press. In this second preface, the author expresses his satisfaction for having been chosen by this particular editor, as Cambridge represents the heart of nuclear physics.  The book was quickly sold out and very hard to find for many years. More recently it has been re-published in paperback by the Cambridge University Press and can still be considered as an excellent example of science popularization.

In 1946, the same year when *Atomic Energy in Cosmic and Human Life* is first published, Gamow delivers to Viking Press another manuscript. It is a very peculiar text, that will be published the following year with a curious title: *One Two Three ...Infinity. Facts and Speculations of Science* [GG47a].

The title itself is quite intriguing. Gamow recalls a passage from Lewis Carroll's book, *Through the Looking Glass.* Quoting from Carroll "The time has come" the Walrus said "to talk of many things", then Gamow continues the sentence as follows: "... of atoms, stars, and nebulae, of entropy and genes; and whether one can bend space, and the rocket shrinks" — where Carroll's original text actually read "... of shoes — and ships — and sealing-wax — of cabbages — and kings — and why the sea is boiling hot — and why the pigs have wings."

With this short introduction, Gamow reveals some of the features his book is about: the physics of microcosm, astrophysics, statistics, biology and the theory of relativity.

The target of this book, in Gamow's intentions, is very ambitious. Quoting the author, "The book originated as an attempt to collect the most interesting facts and theories of modern science in such a way as to give the reader a general picture of the universe in its microscopic and macroscopic manifestations..."

The peculiar choice to link his book to Carroll's, probably indicates Gamow's desire to lead the reader onto paths of science way beyond the usual ones, and into theoretical abstractions. It is a path that somehow leads into a *reflected* dimension, like the one imagined by Lewis Carroll, beyond the looking glass.

The description of the physical world — from microcosm to macrocosm — presented in this book, is very detailed and begins with an interesting overview of a number of mathematical and geometrical aspects. As Gamow points out, some of them help understand the topics treated in the book, whereas others are pure exercises of mental *training*. It is quite clear that Gamow wanted to involve the readers and make them curious to understand and enjoy such complex concepts.

Gamow begins with a question, that is clearly connected to the title of the book "how high can you count?", and immediately after goes on with a more difficult one "how to count *infinities*?". Venturing into geometry, Gamow introduces ideas such as *geometry without measurements of lenghts or angles*, i.e. topology, *one of the most provocative and difficult of the departments of mathematics*, in his own words.

This book had great success, as it offered a thorough overview of the scientific knowledge of the past century and largely contributed to spread an interest for science in the contemporary society. Many physicists admitted to have been stimulated to venture into physics after reading *One Two Three ...Infinity.*

To the same mathematical area, although stylistically very different, belongs another book, *Puzzle Math* [GG56]. Gamow wrote it in 1956 together with Marvin Stern, a mathematician working at Convair, in San Diego, the company Gamow, Bethe and Teller had collaborated with. The book contains about thirty mathematical puzzles, whose solutions are presented in the form of funny short stories.

Part of Gamow's literary production is dedicated to the history of physics, focusing, in particular, on the biographies of some of the greatest physicists. The first book in this specific area is *Biography of Physics, the great physicists from Galileo to Einstein* (1961) [GG61]. It is an excursus along the milestones of the history of physics, from ancient Greece to modern particle physics, through its most renowned representatives. It is the result of an accurate research work, rich of biographic information and historical data.

In *Gravity* (1962) [GG62], Gamow faces the concept of gravity by comparing the theories of the three main figures that investigated it: Galileo Galilei, the first scientist to make "a truly scientific approach to the question of how things fall"; Isaac Newton, the father of the theory of universal gravitation, to whom the largest part of the book is dedicated; Albert Einstein, with his revolutionary interpretation of gravity in terms of the curvature of the relativistic four-dimensional space-time continuum. Accurate explanations, nice portraits of the three scientists, as well as clear explanations, enriched with easy formulas and hand-made sketches, all together make this a very pleasant book to read.

In a similar style, in 1966, Gamow publishes *Thirty years that shook Physics* [GG66]. He goes through the crucial steps of the radical change of perception of the world (in this sense, shocking) represented by quantum physics. Hufbauer in the before-mentioned biographic essay [Hu], reports that Gamow started writing this book in 1962, after two tragic events that had deeply touched him: the car accident occurred to his friend Landau in January, from which he never fully recovered, and the death of his friend and colleague Niels Bohr.  The book represents a tribute to the greatest names in contemporary physics, in order to preserve the memory of their great contributions to the new frame of quantum physics. From Max Planck's very first idea about light quanta, to whom the first chapter is dedicated, the book focuses, a chapter after the other, on the main physicists of the past century: Bohr and quantum orbits; Pauli and the exclusion principle; De

Broglie and pilot waves; Heisenberg and the uncertainty principle; Dirac and antiparticles; Fermi and particle transformations and Yukawa and mesons.

At the beginning of each chapter, every physicist is presented not with a photograph, but with a portrait, hand-drawn by Gamow himself. The book is concluded with a bizarre and goliardic reinterpretation of Goethe's Faust, the *Blegdamsvej Faust*. The setting is the Institute of Theoretical Physics in Copenhagen. Goethe's original characters are substituted by physicists, who are trying to solve a debate (that really had taken place) between Chadwick (the discoverer of the neutron) and Pauli (the inventor of the neutrino), as to the name to be given to their respective particles. Mephistopheles/Pauli tries to persuade Faust/Ehrenfest about the existence of his particle, involving in the discussion the Lord/Bohr, the archangels (Eddington, Jeans and Milne), four strange ladies dressed in grey (gauge invariance, fine structure constant, negative energy and singularity), and of course Gamow, Oppenheimer and Dirac themselves. All this takes place in crowded canteens, during Walpurgis nights.

Despite the great success at an international level, not all Gamow's books were appreciated by his colleagues. Many of them criticized the necessarily "superficial" approach that a popular book must have, by its nature.

Anyhow, in 1965 Gamow's literary production was officially recognized, when he received the *Kalinga Prize for the Popularization of Science,* for his great contributions to the communication of scientific knowledge to a general audience.

---- **14** ----

# Things Hang in the Air

*Weiner: I want to ask a couple of concluding questions, which will make you reflect a little bit. In thinking back over the scientific work you have done, which pieces of work or which sequences of work do you feel were of the most importance at the time, had the greatest impact at the time?*

*Gamow: I don't know. Probably the [genetic] code, I don't know. Of course, in all things first of all the question is: If Einstein were to be killed in an automobile accident several months before he published the theory of relativity, how soon would it appear? Well, probably a year's delay and then somebody else would do it. I mean, things hang in the air so. One can say, my gosh, of course, the potential barrier and then the expanding universe and thermonuclear reaction, explained the solar energy production, the formulas used for the hydrogen bomb calculation. But on the other hand, again, of course, expanding universe is not mine; it's Friedmann, you see, and Lemaître was kind of in between ... [AIP]*

## Moving to Boulder, in the Spirit of the "Young Turks"

Known in the US for being an exceptional lecturer, towards the mid of the 1950s Gamow is invited to give some seminars at the University of Colorado, in Boulder and in Aspen. Gamow is on a temporary leave from GWU, and after a stay in San Diego, in California, for a series of lessons at Convair, he goes directly to Boulder. Charles Weiner provides a short report of Gamow's first stay in Colorado, in one of the final notes of his interview [AIP].

At the end of his first seminar in Boulder, Gamow is invited to a party by some members of the department. He is a very respected scientist in Colorado, and some colleagues express their interest in having him as a professor at their department. Gamow understands this invitation as the result of a desire of renovation at the physics department in Boulder. His thoughts go immediately back to Leningrad, when he, together with his friends Bronstein and Landau — the "Young Turks" — in the early 1930s had started a somewhat revolutionary process of renovation of the department there. And it is most probably in this spirit that Gamow considers very seriously the proposal to move to Boulder.

Besides, those are really hard times for Gamow in his private life. He is about to divorce from his first wife Rho, with whom he has shared life changing experiences, during their many years of marriage. According to Eamon Harper, in his biographic essay on Gamow, the main cause of their problems was Gamow's alcohol abuse, combined with his rather "turbulent" lifestyle [Har2].

The opportunity Colorado is offering him now, would allow him to leave the past and all the Washington memories behind, and move on to a new chapter in his life.

After an interview with the president of the University of Colorado, Gamow officially accepts the offer and decides to move to Boulder. The salary is quite low, but money is not a problem — the royalties from his literary production grant him a good income anyway. In 1956, having quickly arranged all the practical aspects, Gamow becomes Professor of Theoretical Physics at the University of Boulder.

About his new home in Colorado, in a letter to one of his colleagues, he ironizes about his Boulder address as follows[1]:

---

[1] As reported by R. Herman in the section *Reminiscences* of the Proceedings of the George Gamow Symposium [HPA].

GEORGE GAMOW
785 Sixth Street
BOULDER, COLORADO 80302
UNITED STATES
NORTH AMERICA
PLANET EARTH
SOLAR SYSTEM
MILKY WAY
VIRGO CLUSTER
THE UNIVERSE

It is during this troubled period of his life, that Gamow receives from UNESCO the Kalinga Prize for popularization of science. Besides being a source of great personal satisfaction, winning this prize also gives him the possibility to travel around the world, for lectures to be given in India, in Japan and in Australia. Travelling becomes for Gamow a good way to outrun his private problems, and it allows him to engage in one of the things he is most passionate about: sharing with his audience all the interesting, fascinating and exciting aspects of science.

In October 1958, Gamow marries Barbara Perkins. She will be by his side during the last years of his life, trying to convince him to enter a rehabilitation path. In the many letters Gamow sends to Barbara during his journeys around the world, sometimes together with his son Igor, Gamow sounds willing to undertake such a path. However, his overall health is too seriously compromised. Between 1967 and 1968, he is admitted twice to the hospital for cardiovascular surgery.

Despite his bad health, however, Gamow's interest for physics is still very lively and he keeps writing scientific papers. He is now working on an idea suggested by P.A.M Dirac in the 1930s, as to some physical *constants* — notably, the gravitational constant — that might vary in time. But Gamow will never complete this work, In August 1968, he prematurely dies.

Gamow's life came to an end suddenly. His *line of Universe*, as he calls it in his autobiography, was interrupted too early, and he left the world with a feeling that something had been left unaccomplished. Many of his ideas, the results of his outstanding genius, remained somewhat suspended. They were "hanging in the air", waiting for someone to further elaborate upon them. And as it happens for an unfinished painting, Gamow's death left many questions unanswered, as to what incredible contributions he could still have given to the development of 20th century science.

In 1971, a twelve-floor building was built in his honor at the campus of the University of Colorado, in Boulder. It is known as the *Gamow Tower*, and it nowadays hosts the Department of Physics.

Perhaps the best synthesis of all Gamow's extraordinary results, is the one that can be found on the commemorative bronze plaque, placed in his memory at the Department of Physics at the George Washington University:

"George Gamow (1904–1968) is renowned for developing the 'Big Bang Theory' of the universe (1948); explaining nuclear alpha decay by quantum tunnelling (1928); describing, with Edward Teller, spin-induced nuclear beta decay (1936); pioneering the liquid drop model in nuclear physics (1928); introducing the 'Gamow' factor in stellar reaction rates and element formation (1938); modeling red giants, supernovae, and neutron stars (1939); first suggesting how the genetic code might be transcribed (1954); and popularizing science through a long series of books, including the adventures of 'Mr. Tompkins' (1939–1967)."

# References

[AFH] R. A. Alpher, J. W. Follin Jr., and R. Herman: Physical Conditions in the Initial Stages of the Expanding Universe, *Phys. Rev.* 92 (1953) 1347.

[AH] R. d'E. Atkinson and F. G. Houtermans: Transmutation of the Lighter Elements in Stars, *Nature* 123 (1929) 567.

[AH48a] R. A. Alpher and R. Herman: On the Relative Abundance of the Elements, *Phys. Rev.* 74 (1948) 1737.

[AH48b] R. A. Alpher and R. Herman: Evolution of the Universe, *Nature* 162 (1948) 774.

[AH88] R. A. Alpher and R. Herman: Reflections on Early Work on "Big Bang" Cosmology, *Phys. Today*, August 1988, 24.

[AIP] Interview of George Gamow by Charles Weiner on 1968 April 25, Niels Bohr Library & Archives, American Institute of Physics, College Park, MD USA, https://www.aip.org/history-programs/niels-bohr-library/oral-histories/4325.

[AIP1] Interview of Léon Rosenfeld by Charles Weiner on 1968 September 3, Niels Bohr Library & Archives, American Institute of Physics, College Park, MD USA, htpps://www.aip.org/history-programs/niels-bohr-library/oral-histories/4848.

[Br] L. M. Brown: The Idea of the Neutrino, *Phys. Today* (September 1978) 23.

[BW] N. Bohr and J. A. Wheeler: The Mechanism of Nuclear Fission, *Phys. Rev.* 56 (1939) 426.

[Ch] J. Chadwick: Possible Existence of a Neutron, *Nature* 129 (1932) 312.

[COBE] J. C. Mather *et al.*: A Preliminary Measurement of the Cosmic Microwave Background Spectrum by the Cosmic Background Explorer (COBE) Satellite, *Astrophys. J. Lett.* 354 (1990) L37.

[Cri] F. H. C. Crick, The Genetic Code: Yesterday, Today and Tomorrow, *Cold Spring Harbor Symposium*, XXXI (1966).

[CW] J. D. Watson and F. H. C. Crick: Molecular Structure of Nucleic Acids: A Structure for Deoxyribose Nucleic Acid, *Nature* 171 (1953) 737.

[DP] R. H. Dicke, P. J. E. Peebles, P. G. Roll, and D. T. Wilkinson: Cosmic Black-Body Radiation, *Astrophys. J.* 142 (1965) 414.

[Edd] A. S. Eddington: *The Internal Constitution of the Stars*, The University Press, Cambridge, 1926.

[Fe33] E. Fermi: Tentativo di una Teoria dei Raggi Beta, *La Ricerca Scientifica* 4 (1933) 491; Tentativo di una Teoria dei Raggi Beta, *Nuovo Cimento* 11 (1934) 1; Versuch einer Theorie der $\beta$-Strahlen, *Zeit. f. Phys.* 88 (1934) 161.

[GC] R. W. Gurney and E. U. Condon: Wave Mechanics and Radioactive Disintegration, *Nature* 122 (1928) 439.

[GC49] G. Gamow and C. L. Critchfield: *Theory of Atomic Nucleus and Nuclear Energy Sources*, Clarendon Press, Oxford, 1949.

[GG28] G. Gamow: Zur Quantentheorie des Atomkernes, *Zeit. f. Phys.* 51 (1928) 204.

[GG30] G. Gamow: Mass Defect Curve and Nuclear Constitution, *Proc. Roy. Soc. Lond. A*, 126 (1930) 632.

[GG31] G. Gamow: *The Constitution of Atomic Nuclei and Radioactivity*, Clarendon Press, Oxford, 1931.

[GG34] G. Gamow: Nuclear Spin of Radioactive Elements, *Proc. Roy. Soc. A* 146 (1934) 217.

[GG35] G. Gamow: Nuclear Transformations and the Origin of the Chemical Elements, *Ohio J. Sci.* 35 (1935) 406.

[GG37] G. Gamow: *Structure of Atomic Nuclei and Nuclear Transformations*, Clarendon Press, Oxford, 1937.

[GG39] G. Gamow: *Mr. Tompkins in Wonderland*, Cambridge University Press, Cambridge, 1939.

[GG39a] G. Gamow and E. Teller: On the Origins of Great Nebulae, *Phys. Rev.* 55 (1939) 654.

[GG39b] G. Gamow: Evolution of Red Giants, *Phys. Rev.* 55 (1939) 791.

[GG40] G. Gamow: *The Birth and Death of the Sun*, Viking Press, New York, 1940.

[GG44] G. Gamow: *Mr. Tompkins Explores the Atom*, Cambridge University Press, Cambridge, 1944.

[GG46] G. Gamow: Expanding Universe and the Origin of the Elements, *Phys. Rev.* 70 (1946) 572.

[GG47] G. Gamow: *Atomic Energy in Cosmic and Human Life, Fifty Years of Radioactivity*, The Macmillan Company, New York, 1946; Cambridge University Press, Cambridge, 1947 (reprinted in paperback edition in 2011).

[GG47a] G. Gamow: *One, Two, Three...Infinity*, Viking Press, New York, 1947 (and in revised edition in 1961); republished by Dover Publications in 1988.

[GG48a] R. A. Alpher, H. Bethe, and G. Gamow: The Origin of Chemical Elements, *Phys. Rev.* 73 (1948) 803.

[GG48b] G. Gamow: The Origin of Elements and the Separation of Galaxies, *Phys. Rev.* 74 (1948) 505.

[GG48c] R. A. Alpher, R. Herman, and G. Gamow: The Origin of the Chemical Elements, *Phys. Rev.* 73 (1948) 803.

[GG48d] G. Gamow: The Evolution of the Universe, *Nature* 162 (1948) 680.

[GG53] G. Gamow: *Mr. Tompkins learns the Facts of Life*, Cambridge University Press, Cambridge, 1953.

[GG53a] G. Gamow: *The Moon*, Henry Schuman Inc., New York, 1953.

[GG54] G. Gamow: Possible Relation between DNA and the Protein Structure, *Nature* 137 (1954) 318.

[GG55] G. Gamow: On Information Transfer from Nucleic Acids to Proteins, *Det Kongelige Danske Videnskaberne Selskab* 22 (1955).

[GG56] G. Gamow: *Puzzle Math*, Viking Press, New York, 1956.

[GG61] G. Gamow: *Biography of Physics*, Harper & Bros., New York, 1961.

[GG62] G. Gamow: *Gravity*, Doubleday & Co., New York, 1962.

[GG63] G. Gamow: *A Planet called Earth*, Viking Press, New York, 1963.

[GG64] G. Gamow: *A Star Called the Sun*, Viking Press, New York, 1964.

[GG65] G. Gamow: *Mr. Tompkins in Paperback*, Cambridge University Press, Cambridge, 1965.

[GG66] G. Gamow: *Thirty Years that Shook Physics*, Doubleday & Co., New York, 1966.

[GG67] R. A. Alpher, G. Gamow, and R. Herman, Thermal cosmic radiation and the formation of protogalaxies, *Proc. Natl. Acad. Sci.* 58 (1967) 2179.

[GG67a] G. Gamow and M. Yčas: *Mr. Tompkins Inside Himself*, Viking Press, New York, 1967.

[GG70] G. Gamow: *My World Line. An Informal Autobiography*, The Viking Press, New York, 1970.

[GH] G. Gamow and F. G. Houtermans: Zur Quantenmechanik des Radioaktiven Kerns, *Zeit. f. Phys.* 52 (1928) 496.

[GI] G. Gamow and D. Ivanenko: Zur Wellentheorie der Materie, *Zeit. f. Phys.* 39 (1926) 865.

[GL] G. Gamow and L. Landau: Internal Temperature of Stars, *Nature* 132 (1933) 567.

[GS40] G. Gamow and M. Schönberg: The Possible Role of Neutrinos in Stellar Evolution, *Phys. Rev.* 58 (1940) 1117.

[GS41] G. Gamow and M. Schönberg: Neutrino Theory of Stellar Collapse, *Phys. Rev.* 59 (1941) 539.

[GT38] G. Gamow and E. Teller: The Rate of Selective Thermonuclear Reactions, *Phys. Rev.* 53 (1938) 608.

[GT39] G. Gamow and E. Teller: On the Origin of Great Nebulae, *Phys. Rev.* 55 (1939) 654.

[Ha] C. Hayashi: Proton-Neutron Concentration Ratio in the Expanding Universe at the Stages Preceding the Formation of the Elements, *Prog. Theor. Phys.* 5 (1950) 224.

[Har1] E. Harper: Getting a Bang Out of Gamow, *GW Magazine*, Spring 2000, page 14, https://physics.columbian.gwu.edu/getting-bang-out-gamow.

[Har2] E. Harper: George Gamow: Scientific Amateur and Polymath, *Phys. Perspect.* 3 (2001) 335.

[Hew] R. G. Hewlett and O. E. Anderson: *The New World, 1939–1946*, The Pennsylvania State University Press, University Park, Pennsylvania, 1962.

[HPA] E. Harper, W. C. Parker, and G. D. Anderson, Eds.: *The George Gamow Symposium*, Astronomical Society of the Pacific, Conference Series, San Francisco, 1996, http://www.aspbooks.org/a/volumes/table_of_contents/?book_id=241.

[Hu] K. Hufbauer: *George Gamow: A Biographical Memoir*, National Academy of Sciences, Washington, 2009, http://www.nasonline.org/publications/biographical-memoirs/memoir-pdfs/gamow-george.pdf.

[JdL] P. Jordan and R. de Laer Kronig: Movements of the Lower Jaw of Cattle during Mastication, *Nature* Vol. 120, No. 3031 (1927) 807.

[MF] L. Meitner and O. R. Frisch: Disintegration of Uranium by Neutrons: A New Type of Nuclear Reactions, *Nature* 143 (1939) 239.

[Pa] A. Pais: *Inward Bound*, Clarendon Press, Oxford, 1986.

[PW] A. A. Penzias and R. W. Wilson: A Measurement of Excess Antenna Temperature at 4080-Mc/s, *Astrophys. J.* 142 (1965) 419.

[RS] R. Stannard: *The New World of Mr. Tompkins*, Cambridge University Press, Cambridge, 1999.

[Rut] E. Rutherford: Structure of the Radioactive Atom and the Origin of the $\alpha$-Rays, *Phil. Mag.* 4 (1927) 580.

[Rut29] E. Rutherford et al.: Discussion on the Structure of Atomic Nuclei, *Proc. Roy. Soc. Lond. A*, 123(792) (1929) 373.

[Se] E. Segré: Nuclear Physics in Rome, in *Proceedings of the Symposium on "Nuclear Physics in Retrospect"*, R. H. Steuwer, Ed., University of Minnesota Press, Minneapolis, 1979.

[Sh] M. M. Shapiro: George Gamow — An Appreciation, in *Cosmology, Fusion and Other Matters: George Gamow Memorial Volume*, F. Reines, Ed., Colorado Association Press, Boulder, 1972.

[SR] H. D. Smyth: *Atomic Energy for Military Purposes*, Princeton University Press, 1945, https://www.osti.gov/opennet/manhattan-project-history/publications/smyth_report.pdf.

[St] R. H. Stuewer: *The Age of Innocence*, Oxford University Press, Oxford, 2018.

[TS] E. Teller and J. Shoolery: *Memoirs: A Twentieth-Century Journey in Science and Politics*, Perseus Publishing, Cambridge (MA), 2002.

[Wei] S. Weinberg: *The First Three Minutes*, Basic Books, New York, 1977.

[Well] A. Wellerstein, "Manhattan Project", *Encyclopedia of the History of Science* (April 1919), doi:10.34758/swph-yq79.

[WF] J. A. Wheeler interviewed by Kenneth W. Ford, February 4, 1994 (AIP), http://www.aip.org/history-programs/niels-bohr-library/oral-histories/5908-6.

[Wph] J. A. Wheeler: Mechanism of Fission, *Physics Today*, November (1967) 52.

[You] W. Yourgrau: The Cosmos of George Gamow, *New Scientist* 48 (1970) 38.

# Index

Made in the USA
Columbia, SC
21 March 2026

d2349eb8-8399-4c29-b1d1-2edb50e04601R01